INNOVATIVE SITE REMEDIATION TECHNOLOGY

BIOREMEDIATION

One of an Eight-Volume Series

Edited by
William C. Anderson, P.E., DEE
Executive Director, American Academy of Environmental Engineers

1995

Prepared by WASTECH®, a multiorganization cooperative project managed by the American Academy of Environmental Engineers® with grant assistance from the U.S. Environmental Protection Agency, the U.S. Department of Defense, and the U.S. Department of Energy.

The following organizations participated in the preparation and review of this volume:

Air & Waste Management Association
P.O. Box 2861
Pittsburgh, PA 15230

American Society of Civil Engineers
345 East 47th Street
New York, NY 10017

American Academy of Environmental Engineers®
130 Holiday Court, Suite 100
Annapolis, MD 21401

Hazardous Waste Action Coalition
1015 15th Street, N.W., Suite 802
Washington, D.C. 20005

American Institute of Chemical Engineers
345 East 47th Street
New York, NY 10017

Society for Industrial Microbiology
3929 Old Lee Highway, Suite 92A
Fairfax, VA 22030

Water Environment Federation
601 Wythe Street
Alexandria, VA 22314

Library of Congress Cataloging in Publication Data

Innovative site remediation technology/ edited by William C. Anderson
288 p. 15.24 x 22.86cm.
Includes bibliographic references.
Contents: -- [1] Bioremediation -- [2] Chemical treatment -- [3] Soil washing/ soil flushing -- [4] Stabilization/solidification -- [5] Solvent/chemical extraction -- [6] Thermal desorption -- [7] Thermal destruction -- [8] Vacuum vapor extraction
1. Soil remediation. I. Anderson, William, C., 1943- . II. American Academy of Environmental Engineers.
TD878.I55 1995 628.5'5 93-20786
ISBN 1-883767-01-6 (v. 1) ISBN 1-883767-05-9 (v. 5)
ISBN 1-883767-02-4 (v. 2) ISBN 1-883767-06-7 (v. 6)
ISBN 1-883767-03-2 (v. 3) ISBN 1-883767-07-5 (v. 7)
ISBN 1-883767-04-0 (v. 4) ISBN 1-883767-08-3 (v. 8)

The material presented in this publication has been prepared in accordance with generally recognized engineering principles and practices and is for general information only. This information should not be used without first securing competent advice with respect to its suitability for any general or specific application.

The contents of this publication are not intended to be and should not be construed as a standard of the American Academy of Environmental Engineers or of any of the associated organizations mentioned in this publication and are not intended for use as a reference in purchase specifications, contracts, regulations, statutes, or any other legal document.

No reference made in this publication to any specific method, product, process, or service constitutes or implies an endorsement, recommendation, or warranty thereof by the American Academy of Environmental Engineers or any such associated organization.

Neither the American Academy of Environmental Engineers nor any of such associated organizations or authors makes any representation or warranty of any kind, whether express or implied, concerning the accuracy, suitability, or utility of any information published herein and neither the American Academy of Environmental Engineers nor any such associated organization or author shall be responsible for any errors, omissions, or damages arising out of use of this information.

Book design by Lori Imhoff

Printed in the United States of America

CONTRIBUTORS

This monograph was prepared under the supervision of the WASTECH® Steering Committee. The manuscript for the monograph was written by a task group of experts in bioremediation and was, in turn, subjected to two peer reviews. One review was conducted under the auspices of the Steering Committee and the second by professional and technical organizations having substantial interest in the subject.

PRINCIPAL AUTHORS

Calvin H. Ward, Ph.D., MPH, Task Group Chair
Foyt Family Chair of Engineering
Director, Energy and Environmental Systems Institute
Rice University

Raymond C. Loehr, Ph.D., P.E., DEE
Hussein M. Altharthy Centennial Chair
Civil Engineering Department
The University of Texas at Austin

Evan K. Nyer, P.E.
Vice President, Technical Resources
Geraghty & Miller, Inc.

Michael R. Piotrowski, Ph.D.
President
Matrix Remedial Technologies, Inc.

J. Michele Thomas, Ph.D.
Senior Research Associate
Department of Environmental Science and Engineering
Rice University

James C. Spain, Ph.D.
Chief, Environmental Biotechnology
USAF Armstrong Laboratory
Tyndall Air Force Base, Fl.

John T. Wilson, Ph.D.
Processes & Systems Research Division
Robert S. Kerr Environ. Research Lab
U.S. Environmental Protection Agency

Robert D. Norris, Ph.D.
Technical Director, Bioremediation Services
Eckenfelder, Inc.

In addition, the following made worthy contributions to the monograph:

Dolloff F. Bishop
U.S. Environmental Protection Agency

Virginia R. Gordy, Ph.D.
Rice University

Maureen Leavitt
IT Corporation

Harry Tsomides
Rice University

REVIEWERS

The panel that reviewed the monograph under the auspices of the Project Steering Committee was composed of:

George Pierce, Ph.D., Chair
Editor-in-Chief
Society for Industrial Microbiology
Cytec

Richard A. Conway, P.E., DEE
Senior Corporate Fellow
Union Carbide Corporation

Joseph J. Cooney, Ph.D.
Department Chair
Environmental Sciences
University of Massachusetts at Boston

James J. Duffy, Ph.D.
Director of Analytical Services
Oxychem

James J. Ferris, Ph.D.
President, Federal Systems
CH2M Hill Engineering

Diane Saber, Ph.D.
Environmental Services Division
Fluor-Daniel

STEERING COMMITTEE

Frederick G. Pohland, Ph.D., P.E., DEE
Chair
Weidlein Professor of Environmental Engineering
University of Pittsburgh

Domy Adriano, Ph.D.
Professor and Head
Biogeochemical Ecology Division
The University of Georgia
Representing, Soil Science Society of America

William C. Anderson, P.E., DEE
Project Manager
Executive Director
American Academy of Environmental Engineers

Colonel Frederick Boecher
Director of Risk Management
Chemical and Biological Defense Command
U.S. Army
Representing, American Society of Civil Engineers

Paul L. Busch, Ph.D., P.E., DEE
President and CEO
Malcolm Pirnie, Inc.
Representing, American Academy of Environmental Engineers

Richard A. Conway, P.E., DEE
Senior Corporate Fellow
Union Carbide Corporation
Chair, Environmental Engineering Committee
EPA Science Advisory Board

George Coyle
Division of Technical Innovation
Office of Technical Integration
Environmental Education Development
U.S. Department of Energy

Timothy B. Holbrook, P.E.
Engineering Manager
Groundwater Technology, Inc.
Representing, Air & Waste Management Association

Walter W. Kovalick, Jr., Ph.D.
Director, Technology Innovation Office
Office of Solid Waste and Emergency Response
U.S. Environmental Protection Agency

Joseph F. Lagnese, Jr., P.E., DEE
Private Consultant
Representing, Water Environment Federation

Peter B. Lederman, Ph.D., P.E., DEE, P.P.
Center for Env. Engineering & Science
New Jersey Institute of Technology
Representing, American Institute of Chemical Engineers

Raymond C. Loehr, Ph.D., P.E., DEE
H.M. Alharthy Centennial Chair and Professor
Civil Engineering Department
University of Texas

Timothy Oppelt
Director, Risk Reduction Engineering Laboratory
U.S. Environmental Protection Agency

David Patterson
Senior Technical Analyst
Waste Policy Institute
Representing, U.S. Department of Defense

George Pierce, Ph.D.
Editor-in-Chief
Journal of Microbiology
Manager, Bioremediation Technology Development
Cytec Industries
Representing, Society of Industrial Microbiology

Peter W. Tunnicliffe, P.E., DEE
Senior Vice President
Camp Dresser & McKee, Incorporated
Representing, Hazardous Waste Action Coalition

Charles O. Velzy, P.E., DEE
Private Consultant
Representing, American Society of Mechanical Engineers

Walter J. Weber, Jr., Ph.D., P.E., DEE
Earnest Boyce Distinguished Professor
University of Michigan
Representing, Hazardous Waste Research Center

REVIEWING ORGANIZATIONS

The following organizations contributed to the monograph's review and acceptance by the professional community. The review process employed by each organization is described in its acceptance statement. Individual reviewers are, or are not, listed according to the instructions of each organization.

Air & Waste Management Association

The Air & Waste Management Association is a nonprofit technical and educational organization with more than 14,000 members in more than fifty countries. Founded in 1907, the Association provides a neutral forum where all viewpoints of an environmental management issue (technical, scientific, economic, social, political, and public health) receive equal consideration.

This worldwide network represents many disciplines: physical and social sciences, health and medicine, engineering, law, and management. The Association serves its membership by promoting environmental responsibility and providing technical and managerial leadership in the fields of air and waste management. Dedication to these objectives enables the Association to work towards its goal: a cleaner environment.

Qualified reviewers were recruited from the Waste Group of the Technical Council. It was determined that the monograph is technically sound and publication is endorsed.

The reviewers were:

Paul Lear
OH Materials, Inc.

James Donnelly
Davy Environmental

American Institute of Chemical Engineers

The Environmental Division of the American Institute of Chemical Engineers has enlisted its members to review the monograph. Based on that review the Environmental Division endorses the publication of the monograph.

American Society of Civil Engineers

Qualified reviewers were recruited from the Environmental Engineering Division of ASCE and formed a Subcommittee on WASTECH®. The members of the Subcommittee have reviewed the monograph and have determined that it is acceptable for publication.

The reviewers were:

Professor Katherine Banks
Department of Civil Engineering
Kansas State University

Jaakko Puhakka
Department of Civil Engineering
University of Washington

Professor H. David Stensel, Chair
Department of Civil Engineering
University of Washington

Hazardous Waste Action Coalition

The Hazardous Waste Action Coalition (HWAC) is an association dedicated to promoting an understanding of the state of the hazardous waste practice and related business issues. Our member firms are engineering and science firms that employ nearly 75,000 of this country's engineers, scientists, geologists, hydrogeologists, toxicologists, chemists, biologists, and others who solve hazardous waste problems as a professional service. HWAC is pleased to endorse the monograph as technically sound.

The lead reviewer was:

Daniel Bonk
Baker Environmental, Inc.

Water Environment Federation

The Water Environment Federation is a nonprofit educational organization composed of member and affiliated associations throughout the world. Since 1928, the Federation has represented water quality specialists including engineers, scientists, government officials, industrial and municipal treatment plant operators, chemists, students, academic and equipment manufacturers, and distributors.

Qualified reviewers were recruited from the Federation's Hazardous Wastes, Industrial Wastes, and Groundwater Committees. A list of their names, titles, and business affiliations can be found below.

It has been determined that the document is technically sound and publication is endorsed.

The reviewers were:

Terry E. Baxter
Assistant Professor
Civil & Environmental Engineering Department
Northern Arizona University

Richard A. Bell
Project Manager
TRW, Inc.

Edward J. Bouwer
Professor of Environmental Engineering
John Hopkins University

Christopher Englert
Dragun Corporation

Roger Hlavek
Chemical Engineer
Naval Air Warfare Center

Byung R. Kim
Principal Staff Engineer
Ford Research Laboratory

***L. Michael Szendrey**
Vice President & General Manager
Biotechnical Processes, Inc.

George M. Wong-Chong
Manager, Remediation Processes
ICF Kaiser Engineers, Inc.

*WEF lead reviewer

Society for Industrial Microbiology

The Society for Industrial Microbiology (SIM) is a nonprofit professional association dedicated to the advancement of microbiological sciences, especially as they apply to industrial products, biotechnology, materials, and processes. Founded in 1949, SIM promotes the exchange of scientific information through its meetings and publications, and serves as liaison among the specialized fields of microbiology. Membership in the Society is extended to all scientists in the general field of microbiology. Corporate membership is available to industrial companies interested in the aims of the Society.

It has been determined that the document is technically sound and publication is endorsed.

ACKNOWLEDGMENTS

The WASTECH® project was conducted under a cooperative agreement between the American Academy of Environmental Engineers® and the Office of Solid Waste and Emergency Response, U.S. Environmental Protection Agency. The substantial assistance of the staff of the Technology Innovation Office was invaluable.

Financial support was provided by the U.S. Environmental Protection Agency, Department of Defense, Department of Energy, and the American Academy of Environmental Engineers®.

This multiorganization effort involving a large number of diverse professionals and substantial effort in coordinating meetings, facilitating communications, and editing and preparing multiple drafts was made possible by a dedicated staff provided by the American Academy of Environmental Engineers® consisting of:

Paul F. Peters
Assistant Project Manager & Managing Editor

Karen M. Tiemens
Editor

Susan C. Zarriello
Production Manager

J. Sammi Olmo
Project Administrative Manager

Yolanda Y. Moulden
Staff Assistant

I. Patricia Violette
Staff Assistant

TABLE OF CONTENTS

LIST OF TABLES

LIST OF FIGURES

1
INTRODUCTION

This monograph on bioremediation is one of a series of eight on innovative site and waste remediation technologies that are the culmination of a multiorganization effort involving more than 100 experts over a two-year period. It provides the experienced, practicing professional guidance on the application of innovative processes considered ready for full-scale application. Other monographs in this series address chemical treatment, soil washing/soil flushing, stabilization/solidification, solvent/chemical extraction, thermal desorption, thermal destruction, and vacuum vapor extraction.

1.1 Bioremediation

Bioremediation exploits the ability of certain microorganisms — heterotrophic bacteria and fungi — to degrade hazardous organic materials to innocuous materials such as carbon dioxide, methane water, inorganic salts, and biomass. Microorganisms may derive the carbon and energy required for growth through biodegradation of organic contaminants, or, transform more complex, synthetic chemicals through fortuitous cometabolism.

The processes discussed in this monograph fall into two categories: natural bioremediation and enhanced bioremediation. Natural bioremediation, sometimes referred to as intrinsic bioremediation, depends on indigenous microflora to degrade contaminants using only nutrients and electron acceptors available in situ. However, biodegradation rates will be less than optimal if the microbes' nutritional and physiological requirements are not met. Enhanced bioremediation technologies increase biodegradation rates by supplying those nutrients, electron acceptors, or other factors that are rate limiting.

Enhanced bioremediation can be used to degrade contaminants in situ or ex situ. In situ and ex situ processes may be used to treat contaminated liquids, solids, or air. Some examples of in situ processes include land treatment, bioventing, liquid delivery, and air sparging. Ex situ technologies include slurry reactors, land treatment, composting, soil-piles, and biofilters.

1.2 Development of the Monograph

1.2.1 Background

Acting upon its commitment to develop innovative treatment technologies for the remediation of hazardous waste sites and contaminated soils and groundwater, the U.S. Environmental Protection Agency (EPA) established the Technology Innovation Office (TIO) in the Office of Solid Waste and Emergency Response in March, 1990. The mission assigned TIO was to foster greater use of innovative technologies.

In October of that same year, TIO, in conjunction with the National Advisory Council on Environmental Policy and Technology (NACEPT), convened a workshop for representatives of consulting engineering firms, professional societies, research organizations, and state agencies involved in remediation. The workshop focused on defining the barriers that were impeding the application of innovative technologies in site remediation projects. One of the major impediments identified was the lack of reliable data on the performance, design parameters, and costs of innovative processes.

The need for reliable information led TIO to approach the American Academy of Environmental Engineers®. The Academy is a long-standing, multidisciplinary environmental engineering professional society with wide-ranging affiliations with the remediation and waste treatment professional communities. By June 1991, an agreement in principle (later formalized as a Cooperative Agreement) was reached. The Academy would manage a project to develop monographs describing the state of available innovative remediation technologies. Financial support would be provided by the EPA, U.S. Department of Defense (DOD), U.S. Department of Energy

(DOE), and the Academy. The goal of both TIO and the Academy was to develop monographs providing reliable data that would be broadly recognized and accepted by the professional community, thereby eliminating or at least minimizing this impediment to the use of innovative technologies.

The Academy's strategy for achieving the goal was founded on a multiorganization effort, WASTECH® (pronounced Waste Tech), which joined in partnership the Air and Waste Management Association, the American Institute of Chemical Engineers, the American Society of Civil Engineers, the American Society of Mechanical Engineers, the Hazardous Waste Action Coalition, the Society for Industrial Microbiology, and the Water Environment Federation, together with the Academy, EPA, DOD, and DOE. A Steering Committee composed of highly respected representatives of these organizations having expertise in remediation technology formulated the specific project objectives and process for developing the monographs (see page iv for a listing of Steering Committee members).

By the end of 1991, the Steering Committee had organized the Project. Preparation of the monograph began in earnest in January, 1992.

1.2.2 Process

The Steering Committee decided upon the technologies, or technological areas, to be covered by each monograph, the monographs' general scope, and the process for their development and appointed a task group composed of five or more experts to write a manuscript for each monograph. The task groups were appointed with a view to balancing the interests of the groups principally concerned with the application of innovative site and waste remediation technologies — industry, consulting engineers, research, academe, and government (see page iii for a listing of members of the Bioremediation Task Group).

The Steering Committee called upon the task groups to examine and analyze all pertinent information available, within the Project's financial and time constraints. This included, but was not limited to, the comprehensive data on remediation technologies compiled by EPA, the store of information possessed by the task groups' members, that of other experts willing to voluntarily contribute their knowledge, and information supplied by process vendors.

To develop broad, consensus-based monographs, the Steering Committee prescribed a twofold peer review of the first drafts. One review was conducted by the Steering Committee itself, employing panels consisting of two members of the Committee supplemented by at least four other experts (See *Reviewers,* page iii, for the panel that reviewed this monograph). Simultaneous with the Steering Committee's review, each of the professional and technical organizations represented in the Project reviewed those monographs addressing technologies in which it has substantial interest and competence. Aided by a Symposium sponsored by the Academy in October, 1992, persons having interest in the technologies were encouraged to participate in the organizations' review.

Comments resulting from both reviews were considered by the Task Group, appropriate adjustments were made, and a second draft published. The second draft was accepted by the Steering Committee and participating organizations. The statements of the organizations that formally reviewed this monograph are presented under *Reviewing Organizations* on page v.

1.3 Purpose

The purpose of this monograph is to further the use of innovative bioremediation site remediation and waste processing technologies, that is, technologies not commonly applied, where their use can provide better, more cost-effective performance than conventional methods. To this end, the monograph documents the current state of a number of innovative bioremediation technologies.

1.4 Objectives

The monograph's principal objective is to furnish guidance for experienced, practicing professionals and users' project managers. The monograph is intended, therefore, not to be prescriptive, but supportive. It is intended to aid experienced professionals in applying their judgment in deciding whether and how to apply the technologies addressed under the particular circumstances confronted.

In addition, the monograph is intended to inform regulatory agency personnel and the public about the conditions under which the processes it addresses are potentially applicable.

1.5 Scope

The monograph addresses innovative bioremediation technologies that have been sufficiently developed so that they can be used in full-scale applications. It addresses all aspects of the technologies for which sufficient data were available to the Bioremediation Task Group to describe and explain the technologies and assess their effectiveness, limitations, and potential applications. Laboratory- and pilot-scale studies were addressed, as appropriate.

The monograph's primary focus is site remediation and waste treatment. To the extent the information provided can also be applied to production waste streams, it will provide the profession and users this additional benefit. The monograph considers all waste matrices to which bioremediation can be reasonably applied, such as soils, sludges, filter cake, air, and water.

Application of site remediation and waste treatment technology is site specific and involves consideration of a number of matters besides alternative technologies. Among them are the following that are addressed only to the extent that they are essential to understand the applications and limitations of the technologies described:

- site investigations and assessments;
- planning, management, specifications, and procurement;
- regulatory requirements; and
- community acceptance of the technology.

1.6 Limitations

The information presented in this monograph has been prepared in accordance with generally recognized engineering principles and practices and is for general information only. This information should not be used without first securing competent advice with respect to its suitability for any general or specific application.

Readers are cautioned that the information presented is that which was generally available during the period when the monograph was prepared. Development of innovative site remediation and waste treatment technologies is ongoing. Accordingly, postpublication information may amplify, alter, or render obsolete the information about the processes addressed.

This monograph is not intended to be and should not be construed as a standard of any of the organizations associated with the WASTECH® Project; nor does reference in this publication to any specific method, product, process, or service constitute or imply an endorsement, recommendation, or warranty thereof.

1.7 Organization

This monograph and others in the series are organized under a uniform outline intended to facilitate cross reference among them and comparison of the technologies they address. Chapter 2.0, Process Summary, provides an overview of all material presented. Chapter 3.0, Process Identification, provides comprehensive information on the processes addressed. Each process is fully analyzed in turn. The analysis includes a description of the process (what it does and how it does it), its scientific basis, status of development, environmental effects, pre- and posttreatment requirements, health and safety considerations, design data, operational considerations, and comparative cost data to the extent available. Also addressed are process-unique planning and management requirements and process variations.

Chapter 4.0, Potential Applications, Chapter 5.0, Process Evaluation, and Chapter 6.0, Limitations, provide a synthesis of available information and informed judgments on the processes. Each of these chapters addresses the processes in the same order as they are described in Chapter 3.0. Chapter 7.0, Technology Prognosis, addresses the future use of bioremediation and identifies elements of the processes that require further research and demonstration.

2
PROCESS SUMMARY

2.1 Fundamentals/Basic Science

Heterotrophic bacteria and fungi are the primary agents of decomposition of natural organic matter in the biosphere. Some of these microorganisms have the capability to utilize natural organic compounds — such as petroleum hydrocarbons, phenols, cresols, acetone, and cellulosic wastes — as sources of carbon and energy. These and other naturally-occurring compounds can be converted to carbon dioxide, methane, water, microbial biomass, and by-products that are usually less complex than the parent material. This process is a major component of municipal and industrial waste treatment systems.

Microbial communities in soil typically have tremendous potential to degrade a wide range of compounds. The highly successful application of bioremediation to the treatment of gasoline leaks from underground storage tanks is attributable to indigenous microorganisms.

To date, most bioremediation technologies deal with treatment of natural organic compounds. A number of bioremediation strategies are used or are being developed to treat synthetic chemicals, including pesticides, chloroaromatic compounds, polychlorinatedbiphenyls (PCBs), chlorophenols and chlorobenzoates, chloroaliphatic compounds, nitroaromatic compounds, aniline, phthalates, dibenzodioxins and dibenzofurans, methyl t-butyl ether, and metals. Some of these highly substituted compounds as well as some naturally-occurring compounds (polycyclic aromatic hydrocarbons with four or more rings) might not be suitable growth substrates, however, they may be degraded as the result of fortuitous *cometabolism*. These fortuitous reactions stem from the broad substrate specificity of some microbial enzymes. In this monograph, this type of transformation is termed "microbial metabolism of contaminants that are not growth substrates." A variety of enzymatic reactions — oxidation, hydrolysis, reductive dehalogenation, and reduction of nitro groups — catalyze cometabolic processes.

The indigenous microbial community may not have the capability to degrade specific synthetic chemicals of concern at a particular site. If treatability studies show no degradation (or an extended delay before significant degradation is achieved), inoculation with strains known to be capable of degrading the contaminant may be helpful. In a process known as bioaugmentation, microbial strains are added that cannot use the contaminant as a growth substrate but, nevertheless, completely degrade the contaminant. *Bioaugmentation* with strains that cannot use the contaminant as a growth substrate but, nevertheless, completely degrade the contaminant, has proven successful in several laboratory applications while a few field trials have been documented: a novel strain of *Pseudomonas cepacia* has been used to degrade trichloroethylene; *Phanerochaete chrysosporium* biodegrades a wide range of organic compounds with nonspecific extracellular peroxidases; and pentachlorophenol has been treated in soil bioreactors by adding active biomass that has been grown on another substrate.

2.2 Biogeochemistry and Biodegradation

The effect that living organisms have on the geochemistry of the environment is known as *biogeochemistry*. Biogeochemical processes control the global cycling of carbon, nitrogen, phosphorus, and sulfur, as well as some trace elements. Biodegradation of organic contaminants and biotransformation of inorganic contaminants are influenced by a variety of biogeochemical processes and conditions in the environment:

- Oxidation-Reduction Potential - As an environment is converted from oxidizing to reducing conditions, different microbial processes and terminal electron acceptors are used for contaminant biodegradation. In general, biodegradation rates tend to be much lower under reducing conditions than under oxidizing conditions. The change in redox potential can be used as an indirect gauge of biodegradation that may have occurred in a contaminated site.

- Nutrient Cycling and Availability - When assessments of nutrient requirements for contaminated environments are made, the form (availability) of the nutrients in the matrix should be considered. Concentrations of carbon (e.g., total organic carbon, dissolved organic carbon, total petroleum hydrocarbons, etc.), nitrogen, and phosphorus are determined from soil/sediment samples taken from the site. The nutrient amendment is formulated according to the amount of carbon that can be biodegraded when a known supply of electron acceptor is added.

2.3 *Site Characterization Relevant to In Situ Bioremediation*

To develop a remedial plan, several studies should be conducted to define the key characteristics of a site:

- Site Characterization - evaluate the geochemical properties of the site, the quality of the groundwater in contact with the contaminants, the flow properties of the contaminated matrix, and the methods for measuring the contaminant concentration.
- Biofeasibility Evaluation - determine toxicity characteristics of impacted soil/sediment and whether bioremediation processes are already occurring. Laboratory treatability studies may be conducted.

In addition, in situ bioremediation of aquifers involves several engineering considerations — hydrogeological analyses of the contaminated aquifer and chemical analyses of the aquifer sediments and the groundwater.

2.4 *Natural Bioattenuation of Hazardous Organic Compounds in the Subsurface*

Most organic contaminants enter the subsurface as an oily liquid, such as a fuel spill or release of a chlorinated solvent. Groundwater moving through the release dissolves some of the contaminant, which then becomes a plume of groundwater contamination. As the plume moves away from its source, natural biological processes may attenuate the contamination in the groundwater. Several studies have documented the phenomenon of natural bioattenuation, and a mathematical model of aerobic bioremediation in aquifers has been developed.

Aromatic organic compounds, such as the alkylbenzenes, certain simple polyaromatic hydrocarbons (PAHs), and some nitrogen-containing heterocyclic organic compounds can be degraded in groundwater in the absence of oxygen. Anaerobic microorganisms may require months or years to adapt to the contaminant. When oxygen is absent, nitrate, sulfate, carbonate, or iron (III) can serve as the terminal electron acceptor. Halogenated organic compounds can also serve as electron acceptors.

2.5 *Bioremediation Processes*

Contaminated liquid wastes, sludges, surface soils, subsurface sediments, and air are all amenable to bioremediation. In situ processes treat contaminated soil or sediment where they exist; treatment of contaminated material in an aboveground reactor or prepared bed comprise ex situ bioremediation processes.

2.5.1 In Situ Bioremediation Technologies

Some in situ processes — land treatment, bioventing, and, in some cases, air sparging — are appropriate for bioremediation in the unsaturated zone. Other in situ processes — liquid delivery systems and air sparging — are designed for bioremediation in the saturated zone.

2.5.1.1 Land Treatment

In situ land treatment is a managed treatment and disposal technology that involves the 1) application of waste, sludge, or contaminated soil to uncontaminated surface soils at a site and then tilling the applied material into the surface soils or 2) tillage of contaminated surface soils. Land treatment capitalizes on the natural assimilative capacity of the soil to decompose and contain the contaminated material in the surface soil layer.

Site preparation activities vary according to site characteristics and regulatory requirements. Trees and rocks may have to be removed, drainage ditches may be needed to intercept seasonally-high perched water table and runoff, the soil pH may require adjustment, or the site may have to be contoured, terraced, or graded.

The waste material is applied uniformly to the prepared area and tilled into a depth of 15 to 30 cm (6 to 12 in.). Nutrient applications may enhance organic degradation rates initially, but may be unnecessary in subsequent years.

Concentrations of total petroleum hydrocarbons may be reduced 90 to 99%. Specific organic contaminants also are degraded. Compounds with higher aqueous solubilities have relatively shorter half-lives than less soluble compounds. The average half-life of typical oil and grease from refinery wastes and oily sludges ranges from 50 to 150 or more days. Much effort is being expended to define degradation rates for volatile and semivolatile compounds: BTEX — benzene, toluene, ethylbenzene, xylenes, and PAHs.

2.5.1.2 Bioventing

Bioventing is the use of induced air movement through unsaturated soils, with or without nutrient addition, to stimulate indigenous microorganisms to convert organic contaminants, such as petroleum hydrocarbons, to less hazardous substances, especially carbon dioxide and water. Many systems that were designed as in situ vapor recovery systems for physical removal of contaminants were subsequently found to stimulate biodegradation as well.

Some systems use air injection wells alone or in conjunction with air recovery wells. Bioventing has its most direct impact on the unsaturated zone. Nutrients may become available to the bacteria through use of minerals present in the soils, addition of a nutrient-enriched leachate, or in the case of nitrogen, through biological nitrogen fixation. Biodegradation rates are enhanced by high moisture content.

Bioventing system designs incorporate some method of introducing oxygen to the unsaturated soils and may include a means for nutrient addition. The wells are placed in the contaminated zone and screened over some interval, depending on the distribution of the contaminant, the soil type distribution, and whether the surface is covered by an impermeable layer.

The process removes and/or degrades a number of organic contaminants that are biodegradable under aerobic conditions. Volatile, nonbiodegradable constituents can be treated, but the cost of offgas treatment will be increased. All constituents must be present at levels that are not toxic to the microflora.

Where applicable, bioventing has the potential to be a low cost remediation method. Because of the reduced need for offgas treatment and lower volumes of air moved through the soils, bioventing should be less costly than vapor stripping.

2.5.1.3 Air Sparging in the Unsaturated Zone[1]

Air sparging is a means of introducing air into the saturated zone to transfer volatile compounds to the unsaturated zone for biodegradation. In the unsaturated zone, air sparging involves the use of high air injection rates and/or closely-spaced injection wells to transfer volatile compounds to the unsaturated zone faster than they can be biodegraded in the saturated zone. This way, a larger volume of soil and more microorganisms are available to degrade the contaminant than would be available in the saturated zone alone.

Air injection flow rates must be carefully controlled to prevent transfer of volatile constituents to the atmosphere. Some form of vapor recovery and, possibly, groundwater protection may be necessary to prevent or limit losses of contaminants.

2.5.1.4 Liquid Delivery Systems

Liquid delivery systems are used for bioremediation of contamination in the saturated zone. The process is analogous to conventional wastewater treatment in that a terminal electron acceptor and inorganic nutrients are added to enhance contaminant degradation. In contrast to wastewater treatment, which takes place in a bioreactor under controlled conditions, in situ treatment is effected in the subsurface by indigenous microorganisms.

The liquid delivery technique has been used most often at sites contaminated with various types of petroleum hydrocarbons. A process variation

---

1 . See also *Innovative Site Remediation Technology: Vacuum Vapor Extraction*—Ed.

has been tested at field scale to treat chlorinated aliphatic solvents, such as trichloroethylene and less chlorinated ethylenes by simulating methanotrophs with methane and oxygen.

Before subsurface bioremediation is initiated, an extensive site characterization must be conducted. Laboratory treatability assays can be conducted using samples of contaminated sediment but not groundwater. Although treatability assays provide information about biodegradation potential and nutrient amendments that enhance it, extrapolations from the laboratory to the field may not be exact. Biodegradation in situ is limited by the rate at which oxygen is transferred to the contaminant-degrading microorganisms.

The delivery system (consisting of wells or trenches) is designed to circulate adequate amounts of nutrients and oxygen through the zone of contamination to maximize contaminant biodegradation. Injection wells or trenches, through which nutrients and oxygen are added, are placed within or close to the contaminated area. Groundwater extraction wells or trenches may be included. Produced groundwater is extracted, treated aboveground if necessary, and then disposed or amended with nutrients and recirculated. The recirculation system is designed to hydraulically isolate the target area and minimize contaminant migration out of the treatment zone.

Oxygen is provided by sparging with air or pure oxygen or by adding hydrogen peroxide to injected water. Because of the limited solubility of oxygen in water, it is difficult to deliver large quantities of dissolved oxygen to contaminated subsurface environments. Nitrate, sulfate, and salts of iron (III), can act as alternate electron acceptors.

The cost of implementing a liquid delivery system depends on several factors: type, amount, and extent of contamination, sediment characteristics, and the source of oxygen.

2.5.1.5 Air Sparging in the Saturated Zone[2]

Air sparging in the saturated zone serves a two-fold purpose: it provides oxygen, which acts as an electron acceptor for biodegradation in the aquifer and the unsaturated zone, and it physically transfers volatile substances to

2 . See also *Innovative Site Remediation Technology: Vacuum Vapor Extraction*—Ed.

the unsaturated zone for capture by an in situ vapor recovery system. Dissolved oxygen is distributed through the aquifer by movement of air bubbles, diffusion, and groundwater movement.

The degree to which biodegradation occurs (as opposed to physical transfer to the unsaturated zone) depends on the characteristics of the contaminants, the site lithology, and the system design. Only minimal volatilization of heavier hydrocarbon blends can be expected, whereas lighter petroleum hydrocarbons will be both biodegraded and physically removed.

The sparging wells are operated intermittently with relatively long times between air injection periods. This practice minimizes air stripping during a lag period, thereby maximizing the contribution of biodegradation. The off/on schedule can be developed by monitoring dissolved oxygen levels in groundwater and volatile compounds in the unsaturated zone.

2.5.1.6 Potential Applications of In Situ Bioremediation

In general, the effectiveness of in situ technologies is dependent on contaminant biodegradability, contaminant concentration, and soil properties. In situ methods require movement of air or water through undisturbed soils to deliver the electron acceptor and/or nutrients. For this reason, in situ processes are adversely affected when soil/sediment hydraulic conductivities are less than 10^{-4} cm/sec (3.3×10^{-6} ft/sec).

Although liquid delivery was the first in situ bioremedial approach for treating subsurface contamination, air sparging has superseded this method at most sites. Air sparging is less expensive, distributes oxygen across the site more quickly, and is associated with fewer operational problems than the liquid delivery method. Liquid delivery may be applicable in situations where a pump-and-treat system is already in place or where site conditions, such as fractured rock aquifers, aquifers with shallow water tables, or formations with narrow saturated intervals, preclude air sparging. Also, if control of plume migration is mandated, liquid delivery may be advantageous.

Bioventing is most applicable where the depth-to-water exceeds ten feet and the surficial soils do not require treatment or are being treated by another method. Shallower soils and sites with shallower water tables can be treated if the surface is capped. Bioventing systems can also remove nonbiodegradable compounds, as well as contaminants that are difficult to degrade, provided that offgas treatment is included in the process.

2.5.1.7 Limitations of In Situ Bioremediation Technologies

Surface bioremediation methods, such as land treatment, use tillage to aerate the soil. With this form of mixing and aeration, there is the potential for volatile emissions. Whether such emissiions occur and whether thay are a problem requires a site specific determination. Sufficient open space must be available for land application of excavated soils.

Clayey materials with low permeabilities are less amenable to bioventing.

The limitations of air sparging for bioremediation are not fully known at this time. In most cases, vapor and groundwater recovery systems are required. The air may be channeled through permeable layers in heterogeneous soils with clay lenses, gravel stringers, etc., directing oxygen away from the contaminated zone.

Liquid delivery processes are limited by the rate of groundwater recirculation relative to the contamination level. Precipitation of nutrients, iron precipitation, or formation of excess biomass may clog soil pores.

2.5.2 Ex Situ Bioremediation Technologies

Ex situ bioremediation technologies are those in which a waste that has been removed from its point of origin is treated in a closed or open bioreactor. Liquids, solids, and vapor are amenable to ex situ treatment.

Bioreactor design for aerobic treatment must solve two problems. First, the bacteria must be in contact with the contaminants for extended periods of time to complete the biochemical reactions. Secondly, the design must ensure oxygen transfer to the bacteria. Energy requirements for oxygen transfer are usually the main operating cost of a bioreactor, other than manpower costs.

2.5.2.1 Ex Situ Treatment of Contaminated Water

Current designs for biological treatment of contaminated groundwater are based on systems originally designed to treat wastewater. Bioreactors for treating contaminated water can be separated into several main types:

- Suspended-growth reactors - The bacteria are grown in the water and intimately mixed with the organic contaminants in the water. Aerated lagoons or basins fall into this category. Oxygen is supplied with a surface aerator or air diffusers. Minimum residence time is about two days.

- Fixed-film reactors - Bacteria are grown on an inert support medium within the reactor. The contaminated water passes over the attached bacteria and forms a thin water film into which the contaminants and oxygen diffuse. The bacteria degrade the organic contaminants and waste by-products (CO_2, H_2O) diffuse into the water film.
- Submerged fixed-film reactors - In this variation of fixed-film reactors, the support medium is submerged in the water in the reactor tank. The water is in constant contact with the bacterial film, as opposed to passing through in thin water films.
- Reactors based on activated carbon - The combination of powdered activated carbon and active bacteria increases the removal capabilities of the treatment system. The activated carbon adsorbs organic contaminants and acts as an attachment site for bacteria. Another design, a fluidized-bed reactor, is basically a submerged fixed-film system. The support medium consists of small diameter particles. As water and air flow upwards through the medium, the bed is fluidized. Recently, activated carbon has been the main medium used in these systems.
- Miscellaneous designs - Currently, there are several anaerobic reactor designs available. The main applications to date have been in the food and beverage industries for treating wastewaters with high concentrations of organic constituents.

2.5.2.2 Ex Situ Treatment of Contaminated Soils/Sediments

Ex situ biotreatment reactors for soil remediation fall into two main categories: slurry-phase treatment and solid-phase treatment.

Slurry-phase treatment involves maintaining contaminated soils or sludges as an aqueous slurry. Solid-phase biotreatment includes land treatment, soil-pile treatment, and composting. One of the major costs for all slurry and solid-phase reactors is soil/sediment movement. Therefore, the most economical use of bioreactors is when in situ treatment is not feasible and excavation is required.

Either bioreactor vessels or lined lagoons may be used for slurry-phase biotreatment. Basic features include aeration equipment, mechanical mixing, and sometimes, an emissions control system. Soil or sludge is mixed with nutrient-amended water to form an aqueous slurry. The system is operated to maximize mass transfer rates and contact time between the contaminants and microorganisms.

The characteristics of solid-phase biotreatment processes follow:

- Ex situ land treatment - This process is sometimes known as a prepared bed or on-site land-based bioremediation unit. An ex situ unit operates under the same conditions as an in situ land treatment unit with the possible addition of a leachate collection system or liner to prevent migration and loss of contaminants.
- Soil-pile treatment - Two process variations exist: a water-based system that delivers oxygen and nutrients by water movement through the soil, and an aerated soil-pile system in which nutrients are mixed with the soil when the pile is created and oxygen is delivered through air pipes placed in the pile.
- Composting - Composting is a batch biological process used to treat material with high concentrations of biodegradable organic compounds. Waste destruction and conversion are achieved with thermophilic, aerobic microorganisms that occur naturally in decaying organic matter. However, high temperatures (>50°C) rarely are achieved in composting hazardous waste. Typical composting systems are the windrow, in-vessel, and Beltsville.

2.5.2.3 Biological Reactors for Contaminated Air

Biofiltration is the biological removal of organic contaminants from gas streams in a solid-phase reactor. The process is well-established in Europe and Japan as an air pollution control technology.

A biofilter consists of one or more beds of biologically-active material, primarily mixtures of compost, peat, or soil. Contaminated offgas is vented through the filter, and air contaminants diffuse into the wet, biologically active biofilm surrounding the filter particles. Complete biodegradation results in production of CO_2, water, and microbial biomass. Oxidation of reduced sulfur compounds and chlorinated organic compounds also generates inorganic acids.

Biofilters are simple to operate and require little maintenance. The microbes need aerobic conditions and sufficient moisture. The filter medium provides sites for adsorption of pollutants and attachment of microorganisms and also supplies additional nutrients.

2.5.2.4 Potential Applications of Ex Situ Technologies

Use of ex situ technologies may be hampered by buildings or other structures on the site, the depth of the contamination, or location of contaminated soils below the water table. Where excavation is feasible, aboveground methods can be useful.

Contaminants that are more recalcitrant may be more effectively remediated using slurry reactors in which environmental factors can be controlled. For moderately-degradable compounds, particularly if contained in large volumes of clayey soils, low temperature thermal desorption, for volatile and semi-volatile compounds, may be more effective than soil-piles.

Slurry reactors, especially if coupled with soil washing, can treat a wider range of soils and contaminants than most other bioreactors. Liquid reactors are suitable for treating groundwater recovered from pump-and-treat systems; such reactors are particularly effective for treating soluble, nonvolatile, biodegradable organic contaminants.

Air bioreactors are used to treat offgas emissions from bioventing systems, in situ vapor recovery systems, soil reactors, and air stripper towers. They are most applicable if the concentration of volatiles in the air phase is moderate to low.

2.5.2.5 Limitations of Ex Situ Bioremediation Technologies

Soil permeability may limit oxygen transfer in soil-piles during treatment of clayey soils. This problem may be overcome by adding bulking agents or using slow-release oxygen compounds.

Slurry reactors require offgas capture if volatile constituents are present above acceptable levels. This requirement may present a major challenge for lagoons or other large systems. The efficacy of liquid reactor systems may be limited by influent concentrations of organic contaminants that are too high or low for cost-effective biodegradation or the need for constant monitoring.

Use of air bioreactors may be limited by the size of the reactor required to treat the air flow and mass of contaminants in the vapor phase. In some cases, the size requirement of the reactor can be reduced by using another method, such as a catalytic converter, during the first few weeks of operation when the concentration of volatile compounds is greatest. Alternatively, the primary system could be made fully operational over a period of several weeks.

2.6 Process Evaluation

Bioremediation processes depend on a living, biological system that runs twenty-four hours a day, seven days a week. Biochemical processes operate best within optimum ranges of pH and temperature; the bacteria must be supplied with material to enhance reaction rates (usually oxygen and nutrients).

Most systems are designed to degrade the organic contaminants to CO_2, H_2O and biomass, although persistant intermediates are typically produced. Under anaerobic conditions, other by-products may result (anaerobic biodegradation of chlorinated aliphatic solvents produces lower substituted chlorinated hydrocarbons and as final by-products, chloroethane or vinyl chloride). Although such reactions occur naturally in the field, bioremediation design should minimize the possibility that such unacceptable by-products would be formed.

The capital and operating costs of bioremediation depend on the type and quantity of organic compounds present, site conditions, the volume of material to be processed, and the site-specific remediation goals. The main direct costs of bioremediation can be attributed to moving liquid or soils to the reaction zone for the purpose of supplying oxygen to aerobic systems and amending with nutrients.

2.7 Technology Prognosis

The future of bioremediation hinges on several key issues. As these issues are resolved, bioremediation will be used to treat more contaminants under more demanding conditions.

- Regulatory concerns - Although recent changes in regulations remove some restrictions on excavating, bioremediating, and reusing soils on site, other regulations may restrict use of nutrient amendments during in situ treatment. In the future, regulations may also control the addition of genetically-altered microorganisms. Cleanup standards, particularly for total petroleum hydrocarbons, may be overly conservative and difficult, if not impossible, to achieve.
- Engineering considerations - Improvements related to delivery of nutrients and/or electron acceptors will expand the conditions under which bioremediation is cost-effective.
- New bioremediation approaches - White rot fungus, cometabolic processes, anaerobic processes, methods for improving microbial transport through porous media, use of vegetation, and genetically-engineered microbes have been shown to be effective in the laboratory and, in some cases, at pilot scale. Commercialization will require additional effort.
- Additional data - Improvements are needed in methods of site investigations, determination of nutrient and electron acceptor requirements, biotoxicity avoidance, and materials handling. Also needed are better methods for measuring contaminant concentrations and biological activity.

3
PROCESS IDENTIFICATION AND DESCRIPTION

3.1 Overview of Bioremediation

Bioremediation, for the purpose of this monograph, is the process by which organic hazardous materials are biologically degraded, usually to innocuous materials such as carbon dioxide, methane, water, inorganic salts, biomass, and by-products that are less complex than the parent compound. The process is basically an extension of the carbon cycle in which organic and inorganic forms of carbon are cycled back and forth through oxidation and reduction reactions. Figure 3.1 (on page 3.2) depicts the concepts involved in bioremediation of organic contaminants. For bioremediation in situ, biodegradation is effected by the indigenous microflora. Ex situ treatment in bioreactors may entail inoculation with contaminant-degrading microorganisms.

The concept of bioremediation can be traced back centuries to such processes as composting of organic wastes for soil conditioners and mulch (Thomas, Ward et al. 1992). The technology has been expanded to include the treatment of food wastes, agricultural wastes, and wastewater. More recently, bioremediation concepts have been applied in treating hazardous wastes and remediating contaminated soils and groundwater. Enhanced bioremediation was used to treat industrial wastes as early as the 1940s when the petroleum industry managed refinery wastes by land application or treatment, which includes biological, as well as physical and chemical, processes. Management of wastes by land application was unregulated at that time.

Figure 3.1
Bioremedial System

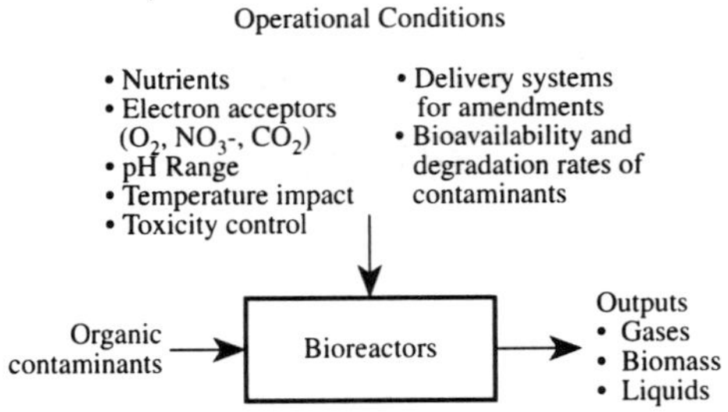

Bioremediation can be used to treat soil, water, and gases that contain biodegradable organic compounds. Although there are many bioremedial processes that can be considered, there are common factors that affect all such processes. These must be understood to obtain the desired performance. The major factors are noted in the above schematic. In addition, there are outputs and residues that may require subsequent management.

Bioremediation must be considered as a system with all of the inputs, outputs, and operational requirements carefully considered. The difference in bioremediation processes, as applied, is due to the media (soil, water, gas) being remediated, the site-specific operational requirements, and any outputs that must be managed.

The types of processes discussed in this monograph include natural and enhanced bioremediation. Natural bioremediation is the process by which contaminants are degraded by indigenous microorganisms that use available nutrients and electron acceptors. As a result of natural bioremediation, many contamination events may go unnoticed. When contamination is detected, the concentrations of contaminants are usually lower than what would be expected from the action of abiotic processes alone. Often, however, the rate of natural bioremediation is limited by nutrient and electron acceptor availability, and the contaminants pose human health and environmental risks. In these instances enhanced bioremediation is applied. Enhanced bioremediation is the process by which the rate of contaminant degradation is increased by adding an electron acceptor, nutrients, or other countering factors that are limiting.

Enhanced bioremediation processes can be used to treat contamination in situ or ex situ (figure 3.2 on page 3.3). In situ processes include treatment of contaminants in the tillage zone (land treatment) and unsaturated and saturated subsurface zones. Land treatment relies on the physical, chemical, and biological processes for attenuation of contamination. Following are the desired results of land treatment:

Figure 3.2
Applicable Bioremediation Processes

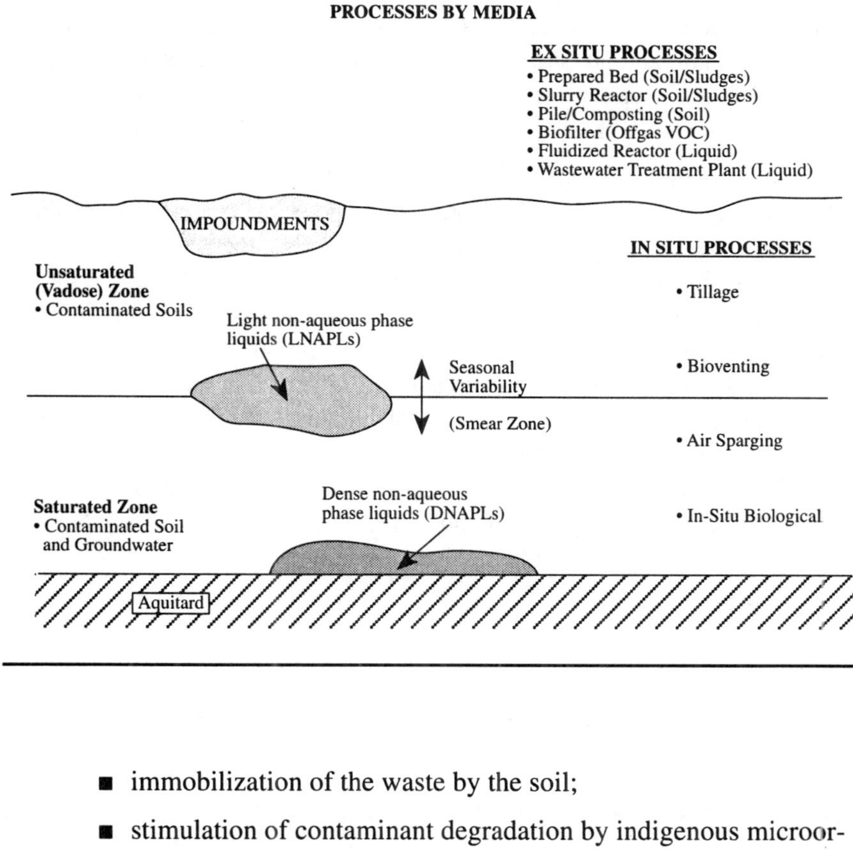

- immobilization of the waste by the soil;
- stimulation of contaminant degradation by indigenous microorganisms or by inoculation;
- minimization of volatilization and leaching of contaminants out of the treatment zone; and
- control of surface water runoff (American Petroleum Institute 1983).

Technologies for treating contamination in the unsaturated and saturated zones include bioventing, air sparging, and biofilters (figure 3.2). *Bioventing* is the process by which contaminants in the unsaturated zone are removed by volatilization and biodegradation as oxygen is supplied by vacuum extraction and/or injection. The key objective in bioventing is to maximize the rate of contaminant removal by volatilization and biodegradation or minimize cost of operation by promoting biodegradation versus volatilization. *Air sparging* entails injecting air into the saturated zone to enhance biodegradation of aquifer contaminants and/or transfer of volatile groundwater contaminants to the unsaturated zone for treatment. By designing systems to deliver limiting nutrients and a terminal electron acceptor in solution, contaminants in the saturated zone can be treated.

Ex situ processes include those for treating contamination in liquids, solids, and air (see figure 3.2 on page 3.3). Processes for treatment of liquids use suspended-growth, fixed-film, and submerged fixed-film bioreactors. Each type of reactor is designed to maximize contact of microorganisms, contaminants, and required nutrients to increase rates of contaminant biodegradation. Solids can be treated using slurry phase processes, land treatment, or composting/soil piles. Each of these processes is managed in a manner designed to maximize rates of biodegradation and/or minimize effluent concentration. *Biofilters* are composed of a solid phase, such as compost, peat, or soil, through which a stream of contaminated air is passed. Vapor-phase contaminants in the air stream are biodegraded aerobically by microorganisms in the solid phase.

3.2 Fundamentals and Basic Science

3.2.1 Microbial Ecology and Physiology

The primary agents of bioremediation are the heterotrophic bacteria and fungi, which derive the carbon and energy required for growth from organic compounds. They are the primary agents of decomposition of natural organic matter in the biosphere and of organic pollutants that resemble the natural substrates of these organisms. Organic contaminants may be degraded by a single type of microorganism, but more likely, by complex mixed cultures, the components of which may require special environmental conditions. It is critical that the biotreatment system be designed with an understanding of the biological processes involved. Understanding metabolic pathways allows evaluation of the extent of biodegradation, intermediate metabolites that might accumulate, and requirements (such as electron acceptors) that must be fulfilled for successful bioremediation.

The range of biodegradative processes available and the complexities of the contaminants must be matched. Microorganisms with specialized metabolic capabilities degrade only particular contaminants; complex microbial communities possessing an array of biodegradative capabilities are site-specific and cannot be considered as generic "biomass." Similarly, contaminants include a wide spectrum of functional groups and chemical properties and should not be considered as "organic loading."

3.2.1.1 Organic Contaminants as Growth Substrates for Microorganisms

The ideal application of bioremediation would be in treating natural organic compounds, such as petroleum hydrocarbons, phenols, cresols, acetone, and cellulosic wastes, all of which can serve as growth substrates for microorganisms (table 3.1 on page 3.6). The biodegradation pathways for most of these compounds have been studied extensively and are well-defined (Gibson 1984). They can be converted to carbon dioxide, water, and microbial biomass with no accumulation of by-products or metabolites. This process, called *mineralization* or complete biodegradation, is a major component of the global carbon cycle. It is a self-sustaining process and requires only the appropriate conditions of temperature, pH, moisture content, inorganic nutrients, and an electron acceptor. The biomass regenerates itself through growth and requires no source of energy other than the organic compound. Such processes have formed the basis of municipal and industrial waste treatment systems for many years. The processes have also been used for bioremediation of creosote in wastes from wood treatment, as well as petroleum hydrocarbons in refinery wastes, oil spills, and subsurface material contaminated by fuels from leaking underground storage tanks.

In recent years an increasing number of synthetic organic (xenobiotic) compounds have entered the environment and become subject to mineralization. A variety of chloroaromatic and nitroaromatic compounds can be readily mineralized if the appropriate microorganisms and conditions are present. Indeed, many pesticides and detergents are now designed to be mineralized in the environment in contrast to earlier products that were very resistant to biodegradation.

Table 3.1
Examples of Naturally-Occurring Organic Compounds Amenable to Bioremediation

Chemical Group	Molecular Formula	
Petroleum Hydrocarbons		
Aliphatic (straight chain)	$C_{16}H_{34}$	(hexadecane)
Aromatic (ring)	C_6H_6	(benzene)
Phenols (hydroxylated ring)	HOC_6H_5	(phenol)
Cresols (methylated phenol)	$HOC_6H_4CH_3$	(*o-*,*m-*, *or p-*cresol)
Acetone (ketone)	CH_3COCH_3	
Cellulose (polysaccharide of repeating glucose units)	$C_6H_{10}O_5$	

3.2.1.2 Requirements for Microbial Growth and Metabolism

Harnessing a natural microbial community for use in bioremediation requires a good working knowledge of its nutritional and environmental requirements. These requirements involve not only the basic parameters such as temperature, pH, and inorganic nutrients common to all biological processes, but also specific requirements for the degradation of a given pollutant by a specific biochemical strategy. A brief discussion of general requirements is provided below and is followed by a more detailed discussion of several degradative processes of potential use in bioremediation.

3.2.1.2.1 Temperature Thermophilic bacteria can be useful for biodegradation at temperatures ranging from 40° to 60°C (104° to 140°F), but so far their use has been primarily limited to composting applications. Biodegradation can also be detected at temperatures as low as 0° to 10°C (32° to 50°F); even though biodegradation rates are slower at low temperatures, successful bioremediation at low temperatures has been reported. Temperatures in the mesophilic range, between 10° and 40°C (50° to 104°F), are more practical for field applications, and in some instances, contaminated materials have been heated artificially to this range as a prerequisite for biotreatment. Within the mesophilic range, microbial growth rates and biodegradation rates increase with increasing temperature. In the prediction of rates, it must be borne in mind that the agents of biodegradation do not constitute a simple catalyst, but a complex community of microorganisms. Thus, a shift in temperature of a few degrees can cause dramatic changes in the composition and function of the community. The specific population, which is able to degrade the contaminant, may function over a narrow range of temperatures or may be replaced by populations with different degradation kinetics or mechanisms. Therefore, the linear relationship that results from a simple Arrhenius plot of reaction rate ($\log_{10}$ k) vs. temperature (1/T) may not apply in many instances and can serve only as a general guideline; prediction of changes in biodegradation rates as a function of temperature must be specific to the particular site and situation. Avoiding temperature fluctuations will improve the effectiveness of bioremediation.

3.2.1.2.2 Nutrients Inorganic nutrients, primarily nitrogen and phosphorus, are essential for all biological processes. Nitrogen can be provided in a variety of forms such as nitrate, ammonium salts, and organic compounds, such as urea. Several forms of inorganic phosphorus have been used. For in situ processes, the choice is usually based on site geochemistry because of potential interactions between phosphate and cations in the water or soil. Treatability studies are usually conducted to determine nutrient requirements, and results are specific to the particular site and process. In early bioremediation projects, carbon loading was determined and nitrogen and phosphorus were provided in ratios known to be optimal for municipal waste treatment systems. Such high nutrient levels are not necessary in bioremediation applications where biomass is not lost from the system. For example, in fixed-film systems (all in situ treatments involve fixed films) most nutrients are recycled and retained in the system. In a well-designed process, much of the organic contamination will be mineralized and biomass accumulation and wasting will be minimized.

3.2.1.2.3 pH For bioremediation processes, the optimum pH will be site- and process-specific and must be determined empirically during feasibility studies; usually few problems are encountered within a pH range of 6 to 8.5. Because many biodegradation processes produce acids or bases, the buffering capacity of the system must be sufficient, or neutralizing agents must be added to maintain an optimum pH range. For example, degradation of chlorinated solvents produces hydrochloric acid, and fermentation of sugars produces organic acids.

3.2.1.2.4 Electron Acceptors Much of the energy for growth of microorganisms is obtained during the transfer of electrons from organic substrates to inorganic electron acceptors. Therefore, appropriate electron acceptors are an absolute requirement for biodegradation, and provision of these electron acceptors often constitutes the greatest challenge in the design of in situ bioremediation systems. The common electron acceptors are carbon dioxide, sulfate, nitrate, and oxygen. In some instances, halogenated organic contaminants can serve as electron acceptors when they are used as substrates for reductive dehalogenation. It is important to note that the electron acceptors listed above are not interchangeable. Although some bacteria can use either oxygen or nitrate as the terminal electron acceptor, the microbial communities are distinctly different for each type of electron acceptor. Their metabolic processes and their potentials for biodegradation of pollutants are also very different. The presence of a given type of community will depend on the availability of the inorganic electron acceptors and on the oxidation-reduction (redox) potential of the system. The redox potential that supports oxygen-based metabolism is higher than that required for nitrate-based systems; sulfate- and carbon dioxide (methanogenesis)-based systems require the lowest redox potentials and are strongly inhibited by oxygen (see figure 3.3 on page 3.9). Microbial communities change naturally as the availability of electron acceptors changes. Bouwer (1992) provides an excellent discussion of the relationship between redox potentials and biodegradation.

An additional consideration in applying oxygen-based systems is the use of molecular oxygen, not only as the terminal electron acceptor in metabolism, but also for initial enzymatic oxidation of organic molecules. This is particularly important in the case of hydrocarbons that are resistant to degradation in the absence of oxygen. A few hydrocarbons such as toluene and xylenes, and many partially-oxidized organic compounds, such as alcohols and acids, can be mineralized under anoxic conditions. The rates of degradation are slower than the corresponding aerobic processes, and the biomass yields are lower because the anaerobic pathways yield less energy. Aliphatic hydrocarbons are not known to be biodegraded under anoxic conditions. The oxygenase enzymes, which insert one or both atoms of molecular oxygen into an organic compound to yield hydroxyl groups, evolved to attack naturally-occurring hydrocarbons. These enzymes can also catalyze the initial oxidation of a variety of xenobiotic compounds, such as many of the chloroaromatic, chloroaliphatic, and nitroaromatic compounds.

Figure 3.3

Oxidation-Reduction Potentials of Microbial Electron Transfer Reactions

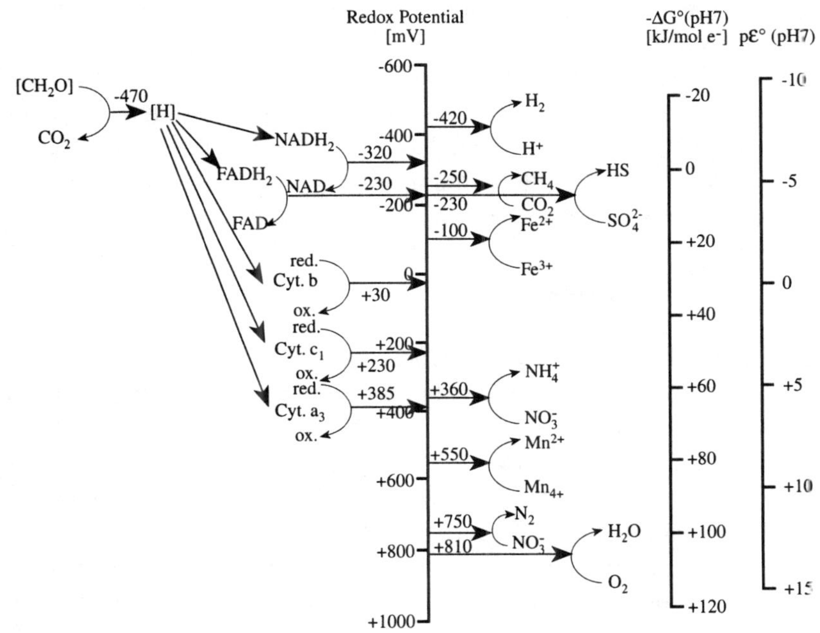

The use of nitrate as a terminal electron acceptor (denitrification and nitrate reduction) shows considerable promise for in situ bioremediation of aromatic hydrocarbons. Denitrification involves the reduction of NO_3^- through the following sequence: NO_3^-, NO_2^-, NO, N_2O, N_2. Nitrate reduction involves the reduction of NO_3^- without the formation of N_2O or N_2. Because nitrate is more soluble than oxygen in water, it may be useful for subsurface bioremediation where transport of oxygen is limited. Low levels of oxygen do not inhibit denitrification, and there is some evidence of synergy between the two processes. Furthermore, heterogeneous micro-environments may exist and allow regions of low redox potential in close proximity to aerobic regions in which denitrification may be inhibited. Therefore, the addition of nitrate might allow conservation of oxygen during treatment, but this possibility has not been rigorously proven.

Reductive dehalogenation reactions, which involve replacement of a halogen with a hydrogen atom, seldom take place in the presence of oxygen, but are favored under anoxic conditions. In some instances, such as with chlorobenzoate, the reaction can be shown to be specific and to yield energy (Mohn and Tiedje 1991). In others, such as reduction of polychlorinated biphenyls (PCBs) or chloroaliphatic compounds, the reactions are poorly understood. The choice of a final electron acceptor dictates the range of potential biotransformation reactions that can be expected in a given system. This concept will be discussed in greater detail, and specific examples will be provided.

3.2.1.2.5 Contaminant Bioavailability Several factors that affect the interactions between organic contaminants and the microbes responsible for their degradation pose major concerns in the design of a bioremediation system. Microorganisms may have the metabolic capability to mineralize a substance and yet fail to do so because it is insoluble, sorbed, or otherwise unavailable to the cell. Biodegradation of polycyclic aromatic hydrocarbons (PAHs) is often limited because the contaminants are not available to the biomass (Stucki and Alexander, 1987). Bioavailability is a particularly important consideration in the design of in situ treatment systems in which the contaminant may be localized as nonaqueous phase liquid (NAPL). Some insoluble hydrocarbons can be degraded in aqueous solutions by microbes in direct contact with the surface of the hydrocarbon, whereas others must be dissolved in the aqueous phase before appreciable biodegradation will occur. Microorganisms often can produce surfactants to aid in the solubilization of poorly soluble or immiscible substrates, and some consideration has been given recently to the use of synthetic surfactants to aid the process. The literature on the degradation of sorbed organic compounds is inconclusive; some studies suggest that sorbed substances must desorb before they are biodegraded, and others suggest that some bacteria can take up certain compounds directly from the sorbed state (Mihelcic et al. 1993). Therefore, contaminant bioavailability must be considered for each site and compound because there is little consensus about which factors influence bioavailability and how bioavailability affects bioremediation.

Two other aspects of substrate availability are also important. Almost all organic contaminants that are candidates for bioremediation are toxic at very high concentrations. But, most of the higher molecular weight hydrocarbons are not toxic at their solubility limits in water. Benzene, however, is more toxic to microorganisms than other hydrocarbons and can only be biodegraded within a certain concentration range. Fortunately, the aqueous concentrations commonly encountered in petroleum-contaminated sites are well below the toxic range. Toxicity might become an important consideration only in sites contaminated with pure benzene. Chlorinated solvents and nitroaromatic compounds also may be toxic at low concentrations. In contrast, PCBs are relatively nontoxic to microorganisms. Toxicity also depends on bioavailability and compounds that are sorbed are often less toxic.

Microorganisms vary widely in their sensitivity to toxic organic compounds; often, a microbial community can adapt and become resistant to high concentrations of a toxic compound. In treatability studies, however, it is very important to determine the toxicity of a waste to the specific organisms able to degrade it and not to the general biomass. Thus, measurement of biodegradation by a microbial community at a variety of contaminant concentrations is a much more important measurement of toxicity than survival or plate counts.

One of the limitations of bioremediation is inherent in the nature of the enzyme reactions involved. The kinetics of enzyme induction and substrate binding and transport dictate that there will be a threshold substrate concentration below which biodegradation rates will be negligible. Contaminants initially present at very low concentrations may not be biodegraded at all. Therefore, concentrations of contamination might be reduced sufficiently by the time biodegradation stops. This must be carefully considered in design of treatment systems and in extrapolation of biodegradation rates measured at one concentration to predict rates at different concentrations.

3.2.1.3 Microbial Metabolism of Contaminants that are not Growth Substrates

Almost all simple organic compounds are susceptible to some type of transformation catalyzed by microbial enzymes. Structural complexity or substitution with halogen, nitro, or other functional groups can produce synthetic chemicals that cannot serve as growth substrates for microorganisms. Some compounds can undergo the initial reactions in a degradative pathway and be converted to intermediates that accumulate because they are not susceptible to attack by subsequent enzymes in the pathway. Such transformations can be useful if the contaminant is rendered less toxic or more bioavailable. Because simple transformations do not yield energy, the microorganisms involved will require an added organic substrate to support growth and activity. Other compounds fail to serve as growth substrates because they do not induce the synthesis of appropriate enzymes. This class of compounds can be degraded if an inducer of the catabolic pathway can be added.

These fortuitous reactions result from the broad substrate specificity of some microbial enzymes and have been called *cometabolism* or cooxidation in the past. But, because such processes can occasionally be shown to yield carbon and energy and convert the contaminant to carbon dioxide (mineralization), cometabolism seems to be less a distinct phenomenon. It is more likely to prevail in natural microbial communities where low concentrations of complex mixtures of organic compounds provide carbon and energy. Growth of microorganisms on high concentrations of a single carbon source is probably a laboratory artifact that bears little relation to natural ecosystems. Because of the many types of transformations that can be considered cometabolism or cooxidation, these reactions have been consolidated under one category of "organic compounds that cannot serve as growth substrates" for the purpose of this monograph. It should also be noted that many synthetic compounds believed to be in this category have recently been shown to serve as growth substrates for specific strains of bacteria. Modern microbial strain construction and selection techniques are rapidly providing new strains with novel capabilities.

3.2.1.3.1 Oxidation The most widely-used microbial transformations involve the oxidation of organic compounds by the introduction of a hydroxyl group derived from molecular oxygen. Such reactions usually are catalyzed by nonspecific monooxygenase and dioxygenase enzymes that insert either one or both atoms of molecular oxygen into the substrate. Such enzymes catalyze the first step in the mineralization of the relatively inert hydrocarbons. Microbial cells grown on hydrocarbons, such as methane or toluene, can oxidize a wide range of organic compounds. An important example is the use of such reactions for the degradation of trichloroethylene, which decomposes spontaneously after the addition of molecular oxygen. Alternatively, hydroxyl groups can be introduced in nonpolar contaminants to increase their solubility, but such an approach must be undertaken cautiously because hydroxylation can increase the toxicity of many contaminants. Oxygenase enzymes can also catalyze the displacement of a wide range of aromatic substituents such as carboxyl, nitro, chloro, ether, and sulfonic moieties.

3.2.1.3.2 Hydrolysis Microbial hydrolases, such as esterases, phosphatases, and lipases, can be used to detoxify or solubilize a variety of contaminants. Because hydrolysis of contaminants provides no energy for the cells, they must be supplied with organic growth substrates. Alternatively, cells can be immobilized and provided with low levels of a growth substrate to support cell maintenance. Because hydrolases are usually stable extracellular enzymes that do not require additional substances for their activity (cofactors), they are also excellent candidates for use as immobilized enzymes. The most likely application of microbial hydrolases will be in treatment of pesticide contamination, because many of the currently available pesticides, such as parathion and diazinon, contain hydrolyzable ester linkages. For such compounds, hydrolysis eliminates or greatly reduces the toxicity.

3.2.1.3.3 Reductive Dehalogenation Replacement of a halogen atom by a proton (reductive dehalogenation) is catalyzed by a variety of anaerobic microbial systems. An example is shown in the following equation:

$$CCl_4 \text{ (carbon tetrachloride)} + 2H \rightarrow CHCl_3 \text{ (chloroform)} + HCl$$

Methanogenic systems, the most widely studied, can dehalogenate contaminants ranging from PCBs to tetrachloroethylene. The reactions seem to be specific for certain isomers in some systems, and in a few cases, the responsible organisms have been studied in pure culture (Fathepure and Boyd 1988). The enzymology of the reactions is poorly understood at present, but recent evidence indicates that the bacteria can derive energy from the use of halogenated organic compounds as an electron sink.

Highly-chlorinated substrates tend to be more susceptible to reductive dehalogenation than their lower chlorinated analogs. Therefore, tetrachloroethylene is dehalogenated more rapidly than trichloroethylene, which is, in turn, more readily attacked than vinyl chloride (Freedman and Gossett 1989). The same sort of relationship applies to the isomeric chlorophenols and the PCBs. This relationship is in direct contrast to that of the oxygenase enzymes mentioned previously, in which more highly-halogenated substrates are less susceptible to attack.

3.2.1.3.4 Reduction of Nitro Groups Nitro groups are the primary functional groups of a variety of pesticides, dyes, and munitions. Many microorganisms synthesize enzymes that catalyze the reduction of nitro groups to the amino level. The reactions with munitions, such as trinitrotoluene (TNT), have been particularly well-studied (McCormick et al. 1978) as have those with dinitrotoluenes (McCormick, Feherry, and Levinson 1976). An example of a reduction of a nitro group follows:

$C_7H_6N_2O_4$ (dinitrotoluene) $\rightarrow$ $C_7H_{10}N_2$ (diaminotoluene)

The reductions can take place under aerobic, as well as anaerobic, conditions and often yield metabolites resistant to further degradation. The reactions studied to date seem to be nonspecific, and the microorganisms responsible require only a source of carbon and energy.

3.2.1.4 Sources of Microorganisms

3.2.1.4.1 Indigenous Microorganisms Microbial communities in soil typically have a tremendous potential for the degradation of a wide range of natural chemicals. Thus, the "natural assimilative capacity" can lead to the degradation of considerable amounts of contamination without any intervention (see Section 3.4 on Natural Assimilative Capacity). This is particularly true for petroleum hydrocarbons; the highly-successful application of bioremediation to the treatment of gasoline from leaking underground storage tanks has been carried out almost exclusively with indigenous microorganisms. During the design of a bioremediation system, it is essential that the presence of appropriate organisms be established. This can typically be done during the treatability study when microbial communities from the contaminated site are tested for their ability to remove the contaminant under laboratory conditions. It is only necessary to measure the *activity* of the microorganisms as shown by the degradation of the contaminant. Indicators of microbial activity include numbers of contaminant-degrading organisms, measurements of contaminant degradation, O_2 uptake, and CO_2 evolution. Excessive amounts of time and money can be spent on enumeration and identification of bacteria in samples from contaminated sites. Although enumeration may be useful as a screening tool, the results seldom, if ever, provide information useful in the subsequent design or operation of the system. Low-microbial activity suggests that some environmental condition is not conducive to rapid contaminant biodegradation and not that the necessary microbes are absent. Occasionally, a low population of specific organisms will produce a low initial degradation rate that will increase after a lag or an acclimation period of a few days. During the acclimation period, the population increases dramatically. If the results of the treatability study indicate that the acclimation period is lengthy, it must be considered in the design of the treatment system.

A wide range of microbial products are marketed for stimulating biodegradation. They are designed to reduce the acclimation period, increase the degradation rate, or increase the range of contaminants removed during bioremediation. There is no evidence in peer-reviewed literature that any of these products has been effective in rigorously-controlled studies. Natural organic compounds, such as petroleum hydrocarbons, are degraded readily by indigenous microbes, and added strains are not likely to compete successfully with the indigenous community. There is substantial interest in the development of specialized microbes, but rigorous experimentation will be required to reveal a successful application.

3.2.1.4.2 Inoculation with Nonindigenous Strains The biodegradation potential of indigenous microbial communities, discussed earlier, applies largely to contamination with natural organic compounds. The potential for degradation of synthetic chemicals may not be present in the indigenous microbial community, particularly if the contamination is recent. Nevertheless, a wide range of synthetic chemicals can be biodegraded, although microorganisms with the appropriate metabolic capabilities may not be widely distributed in the field. This situation would be apparent during treatability studies in which no degradation (or extended delays in the onset of appreciable degradation) reveal the absence of strains able to degrade a contaminant known to be degraded in other systems. In such instances, inoculation with capable strains or starter cultures might prove helpful. Startup of a bioreactor for removal of chloroaromatic or nitroaromatic contaminants, methylene chloride, pesticides, or creosote wastes may benefit from inoculation with such strains. In situ subsurface applications would be less likely to benefit from inoculation because transport of microorganisms is limited in porous media. For inoculation to be useful, it must be demonstrated that the degradative strains compete successfully with the indigenous microbial community during the treatability study.

3.2.1.4.3 Bioaugmentation In some instances, bioremediation can be carried out by nonindigenous microorganisms that cannot use the contaminant as a growth substrate (see Subsection 3.2.1.3 on Microbial Metabolism of Contaminants that are not Growth Substrates). This specific process is called bioaugmentation. The microorganisms must be added continuously or supported on a secondary growth substrate. Delivery and maintenance of the appropriate strains is the most daunting technical challenge in such systems. For example, pentachlorophenol has been treated successfully in soil bioreactors by adding bacteria that completely degrade but do not use the contaminant as substrate (Crawford and Mohn 1985). The system required the regular addition of active biomass grown on another substrate.

Some bioremediation systems for contaminants that cannot serve as a growth substrate will clearly require the addition of active biomass. Thus, the addition of a novel constitutive strain of *Pseudomonas cepacia* capable of degrading trichloroethylene (TCE), shows considerable potential for bioremediation (Shields 1991). Because TCE cannot be used as the sole source of carbon and energy by the *P. cepacia*, periodic renewal of the biomass or addition of a secondary substrate growth will be necessary. Similar considerations may apply to the use of specific methylotrophs, that are able to degrade TCE. Other bioremediation strategies involving cometabolism rely on stimulation of indigenous methylotrophs (Hazen 1991). Several such approaches are in the development stages and may prove useful in the future.

The white rot fungus, *Phanerochaete chrysosporium*, can biodegrade a wide range of organic compounds by the use of nonspecific extracellular peroxidases. It does not use the organic compounds as a source of carbon and energy and must be grown on specific substrates. The lignin peroxidase-mediated biotransformations only occur under conditions of nitrogen limitation and in stationary phase.

Another form of bioaugmentation is the use of immobilized cells or enzymes. Contaminant-degrading organisms are immobilized in or on a solid matrix such as alginate, hollow glass fibers, or porous diatomaceous earth (Stroo 1992). The immobilization confers some protection from toxic levels of contamination and microbial decay. Bioaugmentation has seen limited application for environmental cleanup, but has been very useful in the chemical and pharmaceutical industries. It may provide advantages for bioremediation with genetically-engineered organisms or for treatment of waste streams toxic to microorganisms.

3.2.1.5 Biodegradation of Synthetic Chemicals

Most bioremediation to date has focused on treatment of natural organic compounds, primarily petroleum hydrocarbons. Many synthetic organic compounds can also be biodegraded or transformed by microorganisms, and a number of strategies are being developed to take advantage of these capabilities. This section briefly describes the current stage of development of technologies for bioremediation of representative synthetic chemicals.

3.2.1.5.1 Pesticides For the last twenty years, pesticides have been designed to degrade in the environment, and a considerable amount of information is available on degradation kinetics in soil and water. Less information is available on bioremediation of sites contaminated by accidental releases where concentrations may be much higher than those in routine field applications. Recently, the EPA's Risk Reduction Engineering Laboratory, Cincinnati, Ohio, developed a Pesticide Treatability Data Base that will be useful in evaluating the feasibility of bioremedial applications for treatment of various pesticide wastes.

Most pesticides contain a chemical bond that can be hydrolyzed by microbes or abiotic reactions to yield harmless breakdown products. In some instances, the products can serve as growth substrates for microorganisms, which lead to "acclimation" or selection of a population of specific degraders able to degrade the pesticide at increased rates. For example, carbamates, chlorophenoxyacetates, dinitrocresol, and some organophosphates can serve as growth substrates for soil bacteria. A number of examples of this phenomenon, as well as a good general discussion of pesticide biodegradation, are provided in Racke and Coats (1990). Any pesticide known to serve as a growth substrate for bacteria or stimulate acclimation in soil would be a good candidate for bioremediation.

Other pesticides offer no selective advantage to the microbes that catalyze their degradation. For example, some compounds hydrolyzed by extracellular enzymes yield no degradable products or yield products of no use to the strain that catalyzed the hydrolysis. Other examples are the organohalogen insecticides that are subject to reductive dehalogenation in soil, but provide no advantage to specific microbes. In such situations, acclimation is not observed and degradation rates are often proportional to the total biomass. Recent discoveries that some anaerobes can derive energy from reductive dehalogenation (Mohn and Tiedje 1991) may render this generalization obsolete.

3.2.1.5.2 Chloroaromatic Compounds Biodegradation of chloroaromatic compounds has been studied extensively (Reineke and Knackmuss 1988). Many were thought to be resistant to biodegradation; however, bacteria able to degrade all but the most complex molecules have been discovered during the last twenty years. Chlorobenzenes are the simplest chloroaromatic compounds and, until recently, were not considered to be biodegradable. Bacteria have been isolated that are capable of growing on chlorobenzene (Reineke and Knackmuss 1984; Nishino et al. 1992), 1,4-dichlorobenzene (Schraa et al. 1986; Spain and Nishino 1987), 1,3-dichlorobenzene (deBont et al. 1986), 1,2-dichlorobenzene (Haigler and Spain 1989), 1,2,4-trichlorobenzene (van der Meer et al. 1987), and 1,2,4,5-tetrachlorobenzene (Sander et al. 1991). The metabolic pathways for degradation of the isomeric chlorobenzenes are remarkably similar (Spain 1990) and lead to mineralization with release of the halogens as HCl. Chlorobenzene is the only one of the compounds listed above that has been treated successfully by bioremediation (Nishino et al. 1994). Preliminary results indicate that after long-term contamination, indigenous strains are able to degrade chlorobenzene under aerobic conditions (Nishino et al. 1992). Similar results have been obtained in the laboratory with dichlorobenzenes (Spain unpublished), suggesting that other isomers may exhibit similar biodegradative characteristics. Laboratory and preliminary field work suggest that chlorobenzenes are excellent candidates for bioremediation. Bacteria able to degrade chlorobenzenes may not be ubiquitous; therefore, recently contaminated sites may require extended acclimation periods or inoculation with competent strains. The design of the system must also include the capacity for neutralization of the acid produced during chlorobenzene degradation if contaminant concentrations are high or if a vapor-phase system is used.

3.2.1.5.3 Polychlorobiphenyls (PCBs) Aerobic (Bedard et al. 1987), anaerobic (Brown et al. 1987; Quensen, Tiedje and Boyd 1988), and combination systems (Bedard and Quensen, in press) have been described for the degradation of PCBs. Anaerobic systems are most effective with the more highly-chlorinated congeners whereas aerobic microorganisms work best with the less-chlorinated congeners. No strains with the ability to use complex PCBs as a growth substrate have been isolated. Therefore, bioremediation of PCB-contaminated material is slow and complex. The difficulties are offset, however, by the lack of alternative treatment options for cleaning up large volumes of materials contaminated with low concentrations of PCBs. Pathways initiated by dioxygenase attack seem to be the primary mode of aerobic degradation of PCBs. New bacterial strains are being developed and tested in several laboratories. Their efficacy in the field treatment of PCB-contaminated materials remains to be demonstrated.

3.2.1.5.4 Chlorophenols and Chlorobenzoates Oxidized chloroaromatic compounds, such as the chlorobenzoates and chlorophenols, can often be biodegraded by soil microorganisms. Unfortunately, chlorophenols are toxic to bacteria at high concentrations and can inhibit their own biodegradation. They can be degraded anaerobically via pathways initiated by reductive dehalogenation (Boyd and Shelton 1984; Mikesell and Boyd 1986, 1988). Aerobic degradative pathways are initiated either by hydrolytic removal of a halogen (Apajalahti and Salkinoja-Salonen 1986) or by oxygenase attack on the ring and subsequent removal of the halogen after ring fission (Reineke and Knackmuss 1988).

Although chlorophenols are biodegradable in laboratory studies and in activated sludge, the bacteria responsible are not uniformly distributed. Therefore, recent contamination may require long acclimation periods. Inoculation with adapted strains, as previously noted, may be useful for chlorobenzenes. The toxicity of chlorophenols dictates that the bioremediation system be designed to maintain low concentrations of the contaminant. The release of HCl requires provisions for neutralization of the medium during the treatment of high concentrations.

Pentachlorophenol deserves special mention because of its widespread use as a wood preservative. Aerobic biodegradation by bacteria is initiated by a hydrolytic removal of the *para* chlorine to form tetrachlorohydroquinone (Steiert and Crawford 1986). Subsequent reductive dehalogenations lead to ring fission substrates. Anaerobic degradation proceeds via sequential reductive dehalogenations, leading to mineralization under methanogenic conditions (Mikesell and Boyd 1986, 1988). Pentachlorophenol has been the target of several successful bioremediation demonstrations in the field (Stinson et al. 1991). In the United States, bioaugmentation with a *Flavobacterium* sp. has proven successful (Crawford and Mohn 1985). An alternative process involving mineralization by indigenous strains or a consortium used only as an initial inoculum has also been effective (Compeau, Mahaffey, and Patras 1991). In Europe, an alternative process involving composting has been used (Valo and Salkinoja-Salonen 1986). At wood treatment facilities, the contamination is often a complex mixture including solvents, creosote, metals, and PAHs that can inhibit microorganisms. Therefore, site-specific treatability studies must be conducted for each situation.

3.2.1.5.5 Chloroaliphatic Compounds Chlorinated solvents are among the most common contaminants at waste sites and in groundwater. They are so volatile that they are seldom a problem in soil and surface water. When their volatilization from a contaminated matrix is inhibited, they can be very persistent because they are relatively stable and resist biodegradation. Chloroaliphatic compounds with only one or two chlorine substituents can serve as growth substrates for microorganisms under appropriate conditions. More heavily-chlorinated compounds can only be biodegraded by microbes provided with an alternate growth substrate. Bioremediation in the latter case is much more complex, but the magnitude of the problem and the lack of alternative treatment options have generated a tremendous amount of research on biodegradation of TCE and related solvents. An excellent review on the biochemistry of TCE degradation has been published (Ensley 1991).

Trichloroethylene was considered inert to bacterial degradation until the discovery that cultures grown on methane (methylotrophs) could also oxidize TCE (Wilson and Wilson 1985). The initial step in methane degradation is an oxidation catalyzed by methane monooxygenase. The enzyme is very nonspecific and can also catalyze the oxidation of a wide variety of organic compounds. Subsequent work (Oldenhuis et al. 1989; Tsien et al. 1989) has shown that Type II methylotrophs grown under conditions that select the soluble form of methane monooxygenase have maximum activity toward TCE.

Degradation of TCE by methylotrophs has been studied extensively in the laboratory, at pilot scale, and in field demonstrations (Hazen 1991). A major design difficulty results because methane is a competitive inhibitor of TCE metabolism, and methane monooxygenase has a higher affinity for methane than for TCE. A second problem lies in the irreversible loss of activity of the enzyme during oxidation of TCE, and a third in that the energy required for TCE oxidation must be supplied externally. Methane metabolism can supply the necessary energy, but its competition with TCE makes it a poor choice as an energy source. Formate has been used with some success in the laboratory to provide energy for the process (Oldenhuis et al. 1989).

Trichloroethylene can also be oxidized by ammonia monooxygenase (Arciero et al. 1989), isoprene-oxidizing enzymes (Ewers, Freire-Schroder, and Knackmuss 1990), propane monooxygenase (Wackett et al. 1989), toluene *ortho*-monooxygenase (Shields et al. 1989), toluene *para*-monooxygenase (Winter, Yen, and Ensley 1989), and toluene dioxygenase (Zylstra, Wackett, and Gibson 1989). Each of the other systems mentioned above require the presence of an appropriate inducer that may be a toxic organic compound. Recently, a constitutive mutant of the *Pseudomonas cepacia* strain containing the toluene *ortho*-monooxygenase was developed (Shields 1991). This strain may prove useful, particularly in bioreactors, because it requires no inducer for the toluene monooxygenase.

The products of TCE oxidation depend on the mechanism of the initial oxidation. Monooxygenase attack produces TCE epoxide and chloral (Fox et al. 1990), which decompose spontaneously to dichloroacetate, glyoxylate, formate, and carbon monoxide. In contrast, dioxygenase attack has been reported to initially yield TCE-dioxetane and 1,2-dihydroxy-TCE, which rearrange to formate and glyoxylate (Li and Wackett 1992).

Reductive dehalogenation of chloroalkanes has been reviewed (Mohn and Tiedje 1992). Under anaerobic conditions in the laboratory, TCE, 1,2-dichloroethylene, and vinyl chloride can be converted to ethylene. Initial studies indicated that the reductive dehalogenation required methanogenesis (Freedman and Gossett 1989). Subsequent work has shown that the process can also proceed in the absence of methanogenesis if sufficient methanol is present (DiStefano, Gossett, and Zinder 1991). These studies indicate the potential for bioremediation of this series of chlorinated solvents, but under field conditions the process seldom goes to completion and intermediates, such as vinyl chloride, often accumulate. A number of workers have suggested that reductive dehalogenation might be used for the initial conversion of highly-chlorinated solvents (such as perchloroethylene) to less chlorinated intermediates which could then be degraded oxidatively (see, e.g., Fathepure and Vogel 1991).

Tetrachloromethane can be reductively dehalogenated (Egli et al. 1987), but does not seem to be susceptible to oxidation by bacteria. Similarly, chloroform can be degraded by reductive dehalogenation (Fathepure and Vogel 1991). Dichloromethane can serve as a growth substrate under methanogenic (Freedman and Gossett 1991) or aerobic (Brunner, Staub, and Leisinger 1980) conditions. Therefore, it is an excellent candidate for bioremediation. Pilot-scale studies and field applications will be discussed in a later section.

Dichloroethane can be mineralized under aerobic conditions (van den Wijngaard et al. 1992). Vinyl chloride can also serve as a growth substrate under aerobic conditions for bacteria isolated from soil (Hartmans and deBont 1992), but it is very volatile and would be difficult to treat in a bioreactor.

3.2.1.5.6 Nitroaromatic Compounds Hydroxy- and carboxy-substituted nitroaromatic chemicals are readily degraded by bacteria in soil and activated sludge. They serve as growth substrates for a variety of bacteria, and most of the metabolic pathways have been worked out in detail (Spain and Gibson 1991). The distribution of nitrophenol- and nitrobenzoate-degrading strains in soil and water is often patchy, and extended acclimation periods may be required before rapid biodegradation. Nitrobenzenes and nitrotoluenes are much more resistant to biodegradation; however, nitrobenzene, 4-nitrotoluene, 2,4-dinitrotoluene, and 1,3-dinitrobenzene can serve as growth substrates for bacteria and are good candidates for bioremediation.

Three general mechanisms are used by bacteria for metabolism of aromatic nitro groups. Reduction to the amine level via hydroxylamine seems to be the most widespread mechanism. The resultant amino derivatives are readily degraded, except in the case of the heavily substituted compounds, such as TNT. Partial reduction to the hydroxylamine can lead to subsequent degradation by oxidative metabolism. Nitrobenzene and 4-nitrobenzoate are degraded by this mechanism. Reduction of nitro groups can take place under either aerobic or anaerobic conditions and seems to be a relatively nonspecific process. Oxidative removal of the nitro group was first described for 4-nitrophenol (Spain, Wyss, and Gibson 1979) and 2-nitrophenol (Zeyer and Kearney 1984). The reaction involves the initial replacement of the nitro group with a keto group, then a reduction to the corresponding hydroxy derivative. Dioxygenase enzymes can remove the nitro group from 1,3-dinitrobenzene (Dickel and Knackmuss 1991) or 2,4-dinitrotoluene (Spanggord et al. 1991) and replace it with two adjacent hydroxyl groups. Removal of the nitro group by an oxygenase mechanism typically leads to mineralization of the parent molecule and results in the formation of stoichiometric amounts of nitrite. A third mechanism of enzymatic attack on nitroaromatic compounds has been reported (Lenke, Rieger, and Knackmuss 1992). Reduction of the ring of picric acid by a *Rhodococcus* sp. yielded a hydride-Meisenheimer complex that was subsequently converted to 2,4-dinitrophenol with concomitant removal of one mole of nitrite. This mechanism may eventually be applicable to the degradation of other complex nitroaromatic compounds.

A considerable amount of research has been done on the biodegradation of TNT. Reductive pathways leading to stable amines and dimers were reported in early studies (McCormick, Feeherry, and Levinson 1976). Additional studies have revealed the possibility of more extensive degradation under anaerobic conditions (Roberts, Funk, and Korus 1992). Under aerobic conditions, TNT was shown to be mineralized slowly by *Phanerochaete chrysosporium* (Fernando, Bumpus, and Aust 1990). Several pilot- and field-scale bioremediation projects have been done with systems based on composting (Williams, Ziegenfuss, and Sisk 1992), but little is known about the mechanism of degradation or the final products of the process.

3.2.1.5.7 Aniline Although aniline is not a natural compound, it has been released in the environment for many years because of its extensive use in the chemical industry. It is readily biodegradable under aerobic conditions by a variety of microorganisms. The initial reaction is a dioxygenase-catalyzed removal of the amino group to form catechol, which is the substrate for ring fission. Chloroanilines are more resistant to biodegradation and bind readily to humic material in soil. Humic binding may lead to reductions in bioavailability, mobility, and toxicity of the compound (McCarthy 1989). In aquifer solids and groundwater, chloroanilines are slowly degraded by reductive dehalogenation under anaerobic conditions (Kuhn, Townsend, and Suflita 1990). In pond sediments, an extended acclimation period precedes reductive dehalogenation (Struijs and Rogers 1989). Under aerobic conditions, a *Moraxella* sp. isolated from soil can use chloroanilines as the sole source of carbon and nitrogen (Zeyer, Wasserfallen, and Timmis 1985).

3.2.1.5.8 Phthalates Although phthalates are biodegradable under aerobic conditions in soil and water, extended acclimation periods are often required. They can serve as growth substrates for a variety of bacteria under both aerobic and denitrifying conditions.

3.2.1.5.9 Dibenzodioxins and Dibenzofurans The stability of dibenzodioxins and dibenzofurans is due to the presence of diarylether linkages, which are resistant to enzymatic attack. Tetrachlorodibenzodioxin, coumaphos, and a number of pyrethroid insecticides contain such linkages and are relatively persistent in the environment. Recently, bacteria able to break the ether linkage have been isolated and studied in the laboratory. The degradative mechanism involves angular dioxygenation followed by cleavage of the unstable hemiacetal (Strubel et al. 1991). Dibenzofuran (Fortnagel et al. 1990), coumaphos (Shelton and Somich 1988), and 3-phenoxybenzoate (Topp and Akhtar 1991) can serve as growth substrates for bacteria. A variety of other diphenyl ethers can be metabolized by bacteria grown on dibenzofuran, and several new strains with wider substrate range are under development. Therefore, this class of compounds should be considered as potential candidates for bioremediation. At present the tetrachlorodibenzodioxins remain resistant to biodegradation.

3.2.1.5.10 Methyl t-Butyl Ether (MTBE) The gasoline additive, MTBE, has been used extensively as an octane enhancer. Although there is little evidence that MTBE is biodegraded in the field, a mixed culture has recently been reported to degrade it in the laboratory (Salanitro et al. 1994). The ether does not seem to interfere with degradation of other fuel components.

3.2.1.5.11 Metals Microorganisms can catalyze a wide range of oxidation, reduction, and methylation reactions involving metals. These reactions can result in mobilization, immobilization, or volatilization of the metals. Such reactions are well-documented and show considerable potential, but have been little used in bioremediation because they do not destroy the metals.

Promising efforts are underway on the biological treatment of selenium-contaminated soils (Thompson-Eagle and Frankenberger 1990). There has also been some interest in the use of microorganisms and plants to sequester or concentrate metals from dilute solutions, but the primary importance of metals in bioremediation lies in their toxicity to microorganisms. Heavy metals are used as biocides and can inhibit or kill the bacteria used in biotreatment. Therefore, if they are present at toxic levels in mixed wastes, they must be removed or their toxicity reduced prior to bioremediation.

3.2.2 Biogeochemistry and Biodegradation

3.2.2.1 Introduction

The effect that living organisms have on the geochemistry of the environment is known as *biogeochemistry*. Biogeochemical processes control the global cycling of the biologically-important elements — carbon, nitrogen, phosphorus, and sulfur — as well as the cycling of a variety of trace elements. Biogeochemistry may be influenced by pH, temperature, ionic strength, salinity, and UV light. The processes that influence a number of aspects involved with hazardous materials are also an extension of nutrient cycling by biogeochemical processes. Biogeochemical processes can:

- exert a strong influence on the fate and transport of many toxic organic compounds in a variety of environmental media;
- strongly influence the redox state of a number of hazardous metallic elements, thereby changing their mobility (Moore and Ramamoorthy 1984) and bioavailability (McCarthy 1989); and
- influence the fate and transport of nonmetallic chemical species that pose potential risks to human health and the environment (e.g., nitrate and cyanide)(Keeney 1986; Bulger, Kehew, and Nelson 1989).

In this section, the influences of biogeochemical conditions and processes on biodegradation of organic contaminants and biotransformation of inorganic contaminants in the environment will be explored. Knowledge of the effects of biogeochemistry on biodegradation processes can be used to evaluate relevant site characteristics to assess the feasibility of bioremediation for treatment of a contaminated site.

3.2.2.2 Oxidation-Reduction Potential and Contaminant Biodegradation

Because more energy is derived from aerobic respiration than other microbial processes (figure 3.3 on page 3.9; table 3.2), oxygen is the preferred electron acceptor, if present. The alternate electron acceptors are used more or less sequentially in terms of their energy yields, although some simultaneous usage can occur. Once the concentration of oxygen becomes limiting, between 0.1 and 1.0 mg/L, the microorganisms begin using nitrate and continue using any oxygen that enters the region. Oxidized forms of iron and manganese can serve as electron acceptors in microbial metabolism before sulfate reduction is initiated, which can occur when nitrate is exhausted. Ferric iron, but not manganese, can be used as an electron acceptor in aromatic hydrocarbon degradation. When present, iron and manganese minerals solubilize concurrently with sulfate reduction. The reduced forms of manganese and iron sufficiently scavenge oxygen so that the strict anaerobes, such as some sulfate reducers and all methanogens, can develop.

Many organic compounds (hazardous or otherwise) that enter oxidizing matrices will undergo rapid and extensive biodegradation. This occurs as long as the added substances are not toxic to the receiving microbial community, the microorganisms can readily develop the necessary enzymatic capability to degrade the added substance, and as long as the matrix remains oxidized. In general, biodegradation rates tend to be much lower under

Table 3.2
Microbial Processes and Redox Potential

Microbial Process	Electron Acceptor	Products	Eh (mV)
Aerobic respiration	Oxygen	H_2O	+810
Denitrification	Nitrate, nitrite, nitrous oxide	N_2	+750
Manganese reduction	Mn^{4+}	Mn^{2+}	+396
Iron reduction	Fe^{3+}	Fe^{2+}	-182
Sulfate reduction	SO_4^{2-}	H_2S	-220
Methanogenesis	CO_2	CH_4	-240

Adapted from Bouwer 1984; Schlesinger 1991

reducing than under oxidizing conditions (table 3.3). The difference in the rates can be at least an order of magnitude (Godsy 1987). Furthermore, biodegradation may be incomplete under reducing conditions, in some cases resulting in the formation of metabolic by-products that are more toxic than the parent compound (e.g., the reductive dehalogenation of 1,1-dichloroethene to vinyl chloride under methanogenic conditions (Vogel, Criddle, and McCarty 1987)).

Table 3.3

Degradation of *p*-Cresol Under Various Redox Conditions

Redox Conditions	Biodegradability	Lag Time	Relative Rate	Reference
Aerobic	+		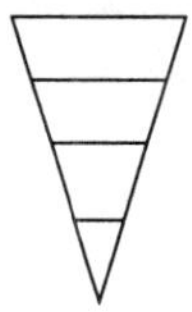	Hopper (1976, 1978)
Denitrifying	+			Bossert and Young (1986)
Sulfate-reducing	+			Bak and Widdel (1986) Smolenski and Suflita (1987)
Methanogenic	+			Smolenski and Suflita (1987) Godsy, Goerlitz, and Ehrlich (1983) Senior and Balba (1984)

From Suflita 1989

The fates of metallic and other inorganic hazardous substances are also influenced by redox potential. Reducing conditions tend to result in a chemical reduction of the introduced substance. For a number of metals, reduced states tend to be more soluble and mobile than oxidized states. Furthermore, studies suggest that if the metallic substances bind to humic materials, both increases and decreases in mobility and toxicity may be observed (for a review, see McCarthy 1989). If nitrate is the contaminant of concern, reducing conditions may lead to nitrate removal (Bulger, Kehew, and Nelson 1989). Oxidizing conditions tend to result in oxidation of the introduced metallic or inorganic substance. The solubility and mobility of a number of metals tend to be reduced once the metal is in a more oxidized form. Inorganic compounds, such as ammonia, can be oxidized (Wilson 1992).

3.2.2.3 Estimates of Biodegradation Using the Expression of Electron Acceptor Demand

The change in redox potential in environments contaminated with organic materials offers a relatively inexpensive tool that can be used as an indirect gauge of the extent of biodegradation that may have occurred in a contaminated site (Wilson, Armstrong, and Rifai 1993). The extent of contaminant biodegradation is estimated by first determining the concentrations of electron acceptors that have been depleted in contaminated groundwater but are present in uncontaminated groundwater very near to the contamination, and the concentrations of ferrous iron and methane which have accumulated in contaminated groundwater. These values are used to estimate the amount of contamination biodegraded using stoichiometry.

The following equations can be used to estimate contaminant biodegradation based on the concentrations of alternate electron acceptors present in groundwater where CH_2 represents a generic alkane, such as pentane or isooctane, and CH represents an aromatic hydrocarbon, such as benzene or toluene:

Aerobic	$4CH_2 + 6O_2 \rightarrow 4CO_2 + 4H_2O$	[1]
Aerobic	$4CH + 5O_2 \rightarrow 4CO_2 + 2H_2O$	[2]
Denitrification	$5CH + 4HNO_3 \rightarrow 4N_2 + 5CO_2 + 2H_2O$	[3]
Sulfate reduction	$8CH + 5H_2SO_4 \rightarrow 4H_2S + 8CO_2 + 4H_2O$	[4]
Iron reduction	$CH + 5Fe^{+3} + 5OH^- \rightarrow 5Fe^{+2} + CO_2 + 3H_2O$	[5]
Methanogenesis	$2CH + 1.5H_2O \rightarrow 0.75CH_4 + 1.25CO_2$	[6]

To assess potential biodegradation in a hypothetical scenario, the following groundwater quality parameters were used: (1) 3 mg/L dissolved oxygen, (2)10 mg/L nitrate as total nitrogen, and (3) 20 mg/L sulfate in uncontaminated groundwater near the impacted area and 40 mg/L ferrous iron and 24 mg/L methane (solubility of methane in cold water) in contaminated groundwater from the impacted area. It is estimated that significant amounts of hydrocarbon can be degraded by natural bioattenuation (table 3.4 on page 3.29). And it is important to note that only oxygen can be used as an electron acceptor in both alkane and aromatic hydrocarbon biodegradation. For aromatic hydrocarbon biodegradation, however, methanogenesis, followed by denitrification, sulfate reduction, and iron reduction, will be more important than oxygen.

Table 3.4

Estimating Biodegradation Using the Concentrations of Alternate Electron Acceptors or Products Found in Uncontaminated and Ground Contaminated Groundwater, Respectively

Equation	Electron Acceptor/Product	Hydrocarbon Consumed
	---------per L groundwater---------	
1	3mg O_2	0.9 mg
2	3mg O_2	1.0 mg
3	17.8 mg HNO_3	5.2 mg
4	20 mg H_2SO_4	4.2 mg
5	40 mg Ferrous iron	1.8 mg
6	24 mg Methane	32.5 mg

3.2.2.4 Nutrient Cycling and Availability

Biogeochemical processes affect nutrient availability and cycling, and, because nutrients are important in biodegradation, they affect contaminant biodegradability. Elements are being transformed constantly, taken up and excreted by organisms and cycled between sequestered and bioavailable forms. When assessments of nutrient requirements for contaminated environments are made, the form (availability) of the nutrients in the matrix should be considered.

An evaluation of nutrient relationships is usually conducted by collecting soil or sediment samples and determining the concentrations of the various forms of carbon (total organic carbon, dissolved organic carbon, and total petroleum hydrocarbons), nitrogen (typically ammonia, nitrate, and nitrite), and phosphorus (typically ortho-phosphate) using standard chemical procedures and protocols. The nutrient amendment is determined first by calculating the amount of carbon that can be biodegraded when a known supply of electron acceptor is added. Using this value, the amounts of nitrogen and phosphorus required to degrade the carbon can be calculated using the carbon:nitrogen:phosphorus (Redfield) ratios found in biological materials. A common mistake made in bioremedial operations is to overestimate nutrient requirements by using the amount of available carbon rather than the amount of carbon that can be consumed, based on the supply of electron acceptor. Addition of unnecessary nutrients is a waste of capital and time.

Ambient levels of nutrients and nutrient cycling from the biomass to the water are considered when the nutrient amendment is formulated. The concentrations of carbon, nitrogen, and phosphorus in samples of water or soil provides little insight into the dynamic nature of nutrient-contaminant relationships, which will be especially important in environments such as a flowing groundwater or streams. Nitrogen and phosphorus may be cycling constantly between the sequestered and bioavailable forms, and in a dynamic site, may be constantly delivered and removed. Furthermore, because measurements are usually made of the most biologically-available forms of nitrogen (ammonia, nitrate, nitrite) and phosphorus (ortho-phosphate), incorrect conclusions can be drawn regarding nutrient balance and availability because their concentrations are often below detectable limits.

Where nitrogen and phosphorus concentrations are below detection limits, it is usually assumed that these nutrients are limiting. This assumption may not be true, since organic forms of nitrogen, that is, those measured by the total Kjeldahl nitrogen method, and phosphorus, as measured by the total phosphorus method and the total reactive phosphorus method, etc., may not be readily available to the indigenous microorganisms, but, nonetheless, may represent important resource pools that can provide the microbes with the necessary nitrogen and phosphorus. These tests may estimate supplies of these nutrients, but their predictive ability is unclear. Indeed, there is evidence that additions of nutrients may actually impede contaminant biodegradation rates in some cases (Cunningham 1992). Because of the dynamic nature of the soils and groundwater, however, only estimates of nutrient concentrations in soils and water can be determined and these values are used to determine nutrient formulations.

3.3 *Site Characterization Relative to In Situ Bioremediation*

3.3.1 Introduction

An important determinant for successful application of bioremediation is the physical nature of the contaminated matrix. Site features strongly influence whether bioremediation is feasible and which bioremediation approach(es) will be effective.

Bioremediation is effective in matrices with high fluid-flow properties. In the saturated zone, matrices with hydraulic conductivities $>10^{-4}$ cm/sec are most amenable to bioremediation through liquid delivery methods. (See Subsection 3.5.1.2.1). In the unsaturated zone, matrices with hydraulic conductivities as low as 10^{-5} to 10^{-6} cm/sec may be treated successfully through bioventing (see Subsection 3.5.1.1.2), which can transport more oxygen per unit volume than liquid delivery. But, these fluid-flow values for treatability are not absolute; the higher the degree of contamination, the greater the limitations imposed by fluid-flow properties.

Where this parameter is lower, the matrix may be more difficult to treat. Matrices having high hydraulic conductivity can receive more air or liquid flow per unit time than matrices with low values. Because air and liquids contain the electron acceptor and possible nutrient amendments, longer treatment will be required for matrices with low-flow properties than for those with high-flow properties.

Matrices having low-hydraulic conductivity often contain clay. Contaminated clays are among the most difficult matrices to treat through bioremediation because (1) contaminants sorb more strongly to clays than sands and (2) the air or liquid flow rate through clays is lower than that for sands. In addition to slow delivery of the electron acceptor and nutrients in clays, bioavailability of the contaminant can be decreased by sorption to clay particles (see Subsection 3.2.1.2.5 Contaminant Bioavailability). One well-understood mechanism that can exclude contaminants from biodegradation is the sorption of organic compounds to the 2:1 layered silicate clays (Stotzky 1986).

A subsurface characteristic that will greatly impact the performance of any in situ bioremediation approach is site heterogeneity. Although the overall hydraulic conductivity of a site may render it amenable to in situ treatment, contamination in zones of low hydraulic conductivity will be more difficult to treat. Air or water that is injected into the subsurface may not fully penetrate or may bypass these zones, leaving pockets of contamination; however, treatment with air in the unsaturated zone will be less hindered than treatment with water in the saturated zone. Further complicating the treatment of heterogeneous sites is the fact that low permeability zones generally have a relatively high capacity to sorb organic contaminants. High sorption capacity results in higher concentrations of contaminants that are more difficult to treat than the contamination found in high permeability zones. The resulting uneven treatment will make it difficult to assess remedial progress.

In some cases, a remedial system can be designed to overcome the problems associated with heterogeneous sites. If contamination has impacted underlying geologic strata with varying permeabilities, it may be possible to place injection points at appropriate depths to access the various affected strata.

Because more permeable strata would not require as high pressures to achieve appreciable injection rates compared to lower permeable strata, it would be necessary to have separate injection systems for permeable versus poorly permeable strata. In this way, the low-permeability strata system can be pressurized appropriately.

The use of separate injection systems is better suited for treatment of horizontal layers of different permeability rather than discontinous lenses. Multiple discontinuous lenses will be the most difficult and expensive to treat. In other cases, once the more permeable zone has been treated below the site specific clean-up goal, the residual contamination in the lower permeable zone may diffuse into the groundwater at a slow enough rate that natural bioremediation (Section 3.4 Natural Bioattenuation of Hazardous Organic Compounds in the Subsurface) may be sufficient to prevent migration.

3.3.2 Overview of a Biofeasibility Assessment Procedure

The key characteristics of a site that must be considered during the development of a remedial plan are summarized in this section. This overview provides recommendations for studies that should be conducted during the remedial investigation (RI) phase.

3.3.2.1 Site Characterization

The first task in site characterization is to acquire a thorough understanding of the three-dimensional distribution of contaminant mass and the distribution of properties at a site that will affect the rate and extent of remediation. This task is important because the site condition is the ultimate determinant of which treatment technologies can be effectively applied. A review of existing site assessment and investigation reports, reports discussing local and regional geology, and interviews with site personnel may be helpful. By way of delineating the context of the contamination, the following should be determined:

- the geochemical properties at the site;
- the groundwater quality in contact with the contaminants;
- the fluid-flow properties of the contaminated matrix; and
- methods for measuring the concentration of contaminants.

The geochemical properties include soil, sediment and rock types; pH (EPA Method 9045); nutrient concentrations (total Kjeldahl nitrogen, EPA Method 351.2; total phosphorus, EPA Method 6010); and total organic carbon concentration (American Society of Agronomy Method 90-2). Knowledge of these characteristics is important because materials added to or produced in the matrix during bioremediation can react with matrix constituents. For instance, geological materials consisting of limestone ($CaCO_3$) or dolomite ($MgCO_3$ and $CaCO_3$) naturally buffer the CO_2 produced during biodegradation, thus maintaining a neutral pH in the treatment zone. These materials also may precipitate some of the added nutrients as salts, which may limit fluid transport through the treatment area. In contrast, matrices composed of highly-weathered materials, which contain little or no calcium and magnesium, will not be able to buffer the CO_2 produced during biodegradation. In highly-weathered materials, pH values can reach levels as low as 4. In addition, materials consisting of clay-sized particles tend to trap oily-phase material, particularly above the water table.

The quality of the groundwater in contact with the contaminants is important. The groundwater contains the nutrients and electron acceptors that will be available to the microorganisms and affects the buffering capacity of the treatment zone. Hard water, which contains high concentrations of cations, will have a high capacity to buffer the CO_2 produced during bioremediation and maintain a neutral pH in the treatment zone; however, these cations may precipitate out with added nutrients, reducing fluid flow. In aquifers with soft water, pH values may be low because the buffering capacity is low.

The fluid-flow properties of a matrix will affect the transport of air or water through it. The fluid-flow properties in the unsaturated and saturated zones are determined by measuring the pneumatic and hydraulic conductivity, respectively.

The horizontal hydraulic conductivity can be measured using a slug, pump, or tracer test (Thomas, Marlow et al. 1992). The type of test will depend on the economics and the information desired. While slug tests are relatively inexpensive, the hydraulic conductivity within only a short radius of the well is obtained. Pump and tracer tests are expensive, but provide a more extensive regional estimate of fluid flow. While pump and tracer tests can be used in any matrix, the slug test is used in those matrices that are moderately permeable-to-tight. The vertical hydraulic conductivity can be determined using two wells placed in the same location where one is screened above the other. A tracer is added to one well and its breakthrough into the second well is measured.

The pneumatic conductivity is measured using two wells. Air is injected into one well and extracted from the second (Cho and DiGiulio 1992). After stabilizing the well pressure, the soil-air distribution between the wells is determined.

Fluid-flow properties should be determined at multiple locations because of the inherent heterogeneity of subsurface formations. Variations in vertical and horizontal hydraulic conductivity can be better defined by determining the vertical stratigraphy at various locations. The stratigraphy can be determined by examination of corings during well installation.

The method used to measure the contamination is also important. For many types of contamination, the bulk of the material is trapped as an oily phase, and only small amounts are dissolved. If based on analysis of groundwater only and not analyses of soil and sediments, estimates of contamination will be low. Borings are made through the contaminated to the uncontaminated material in the contaminant source area, (e.g., underground tank, pipeline rupture) and representative samples are collected at depths that exhibit signs of contamination. Color changes in the samples can be used as indicators of contamination when compared to uncontaminated samples of the same soil type and from the same formation (material contaminated with petroleum compounds will be blue/gray or green, whereas uncontaminated material will be light brown to red). Through analyses of samples of soil or sediment, the following determinations should be made:

- the concentrations of lighter-than-water contaminants (LNAPLs) in the "smear" zone at the fluctuating interface between the vadose and saturated zones beneath and downgradient from the contaminant source;
- the concentrations of sorbed heavier-than-water contaminants (DNAPLs) at intervals across the vertical extent of the affected aquifer beneath and downgradient from the contaminant source; and
- the concentrations of contaminants in the vadose zone.

3.3.2.2 Biotreatability Evaluation

The first step in the evaluation is to determine whether or not microbial processes that will be used during remediation are already occurring at some limited rate (natural bioattenuation). This evaluation can be made by conducting an in situ soil-gas survey in the unsaturated zone and/or by determining groundwater quality. Depending on the requirements of the situation, these tests may be used instead of a laboratory biotreatability study.

For the soil-gas survey, soil-gas samples from the contaminated and uncontaminated zones are collected and analyzed for oxygen and CO_2. Concentrations of oxygen of less than 21% (atmospheric is ≈21%) and CO_2 greater than 1% (atmospheric is ≈0.03%) in the affected zone would suggest contaminant biodegradation, oxygen consumption, and CO_2 production by the indigenous microorganisms. The CO_2 may be produced as a result of aerobic and anaerobic processes. For groundwater, the presence or absence of the electron acceptors (oxygen, nitrate, nitrite, and sulfate) in contaminated and uncontaminated samples may indicate that biodegradation is occurring (Piotrowski 1989). In uncontaminated groundwater, there is usually >1 mg/L O_2, nitrogen is present as nitrate and not ammonia, sulfate is present, and iron and methane are absent. The presence of reduced iron, >0.1 mg/L methane, H_2S, and <0.5 mg/L O_2 in contaminated water would suggest natural bioattenuation is occurring. Groundwater is sampled in a series of wells upgradient and downgradient of the spill. At many sites, existing monitoring wells can be used.

Care must be exercised in measuring O_2 in groundwater. Groundwater containing a layer of free-floating product should not be sampled, to avoid fouling the dissolved oxygen probe. Sampling procedures, such as bailing or pumping a well, can easily reaerate the water. The method chosen for sampling groundwater from a well will be specific to that well and depend on the depth of the water table and the hydraulic conductivity of the aquifer. An indirect indicator of the absence of O_2 would be the presence of reduced iron and/or methane.

Laboratory treatability studies may be conducted. To determine the rate and extent of contaminant biodegradation, samples of sediment or soil are subjected to batch studies using varying nutrient amendments based on the site characterization. However, the fact that laboratory-derived biodegradation data may not be the same as those in the field must be considered. The results of these studies are used to determine the final nutrient formulation required to achieve the most rapid and cost-effective rate of remediation. In addition, these studies can be used to determine whether the prospective nutrient amendment is compatible with the subsurface material. Care must be taken to avoid precipitating the nutrients and plugging the aquifer. In materials containing clay, potassium rather than sodium salts are used to prevent swelling of the clay. When the groundwater is hard, tripolyphosphate instead of orthophosphate is used, because it will solubilize rather than precipitate iron, calcium, and magnesium. In aquifers high in iron, added oxygen can oxidize and thus precipitate the iron.

Although microbial numbers and types have been determined in early bioremediation projects (Raymond et al. 1975; 1978), they are not necessary to predict the feasibility of bioremediation. Biofeasibility tests should be conducted to measure microbial activity (biodegradation) rather than microbial numbers.

3.3.3 Design Considerations for In Situ Bioremediation of Aquifers

Regulatory concerns and site characteristics must be considered during the development of the remedial design. Regulatory concerns are the driving mechanism for aquifer restoration and often have a marked impact on the bioremedial treatment ultimately approved for the site. Failure to address regulatory concerns early in the remedial design process often results in regulatory resistance or outright rejection. The geology, hydrogeology, geochemistry, and biogeochemistry of the site will also affect system design.

3.3.3.1 Regulatory Concerns

Among the most important site aspects that can strongly influence the design of an in situ bioremdiation system is the the regulatory concern regarding the contaminated groundwater. A contaminated groundwater plume located upgradient from a municipal drinking water source or sensitive aquatic habitat will elicit more concern than a plume located in an uninhabited, less-sensitive area.

Sites Upgradient from Sensitive Areas. Concerns over plume migration toward a sensitive area typically result in a regulatory mandate that plume control be adopted as a part of the remedial strategy. Consequently, some form of pump-and-treat system is installed as a hedge against further plume migration.

The aboveground treatment for the extracted groundwater must reduce contaminant levels to concentrations that meet regulatory standards. Often, some form of "rough" treatment is used to cost-effectively reduce the contaminant concentration and then the partially-treated groundwater is passed through a "polishing" step (such as activated carbon) to consistently produce groundwater containing acceptable contaminant concentrations. Aboveground biotreatment options (such as those discussed in Sections 2.5.2.1 and 3.5.2.1) are now commonly used for the rough treatment step.

Historically, the treated groundwater has been considered waste that must be disposed. Generally, the water is discharged to a body of water under a National Pollutant Discharge Elimination System (NPDES) Permit or to a publically-owned treatment works (POTW) under a local permit. In both cases, the pollutant concentrations must not exceed regulatory standards.

Instead of disposing of the treated groundwater, all or a portion of the water can be amended with biostimulating additives and returned to the treatment area as described in Sections 2.5.2.1 and 3.5.2.1, Liquid Delivery. Moreover, if the aboveground treatment process includes a biologically-based "roughing" step, biological "preconditioning" of the water may provide an added stimulating effect by supplying nutrients and microorganisms adapted to degrade the contaminants. However, the transport of microorganisms for the purpose of contaminant biodegradation in the subsurface has not been demonstrated.

As long as the groundwater extraction rate is equal to or greater than the liquid delivery rate, the hydraulic control of plume migration should be maintained (closed loop). McCarty et al. (1989) estimated that coupling a pump-and-treat system with an in situ bioremediation program can reduce the overall time for aquifer restoration by as much as one-half or more.

Assuming re-injection of the treated and amended groundwater is approved, a regulatory constraint may be converted into a substantial remedial benefit. More and more states are recognizing the potential benefits of allowing re-injection of treated groundwater for aquifer biorestoration.

States vary in their regulatory stances towards injection of treated groundwater. A number of states will not allow it at all. In these cases, injection of appropriately amended clean groundwater, surface water, or dechlorinated municipal water may be allowed for in situ bioremediation.

In other states (e.g. Texas), treated groundwater can be used for re-injection as long as the treatment process consistently produces water containing contaminant constituents below applicable regulatory goals. Finally, in Michigan, re-injection of untreated groundwater has been permitted in a number of cases. In most situations involving groundwater reinjection, it will likely be necessary to ensure that hydraulic control of plume migration will be maintained during re-injection so that off-site plume migration does not occur.

An interesting development in this regard was a study conducted at a site in Michigan by a partnership (termed "CoBioReM") involving a state regulatory agency, the oil and gas industry, and two universities in a 1991 study. Contaminated groundwater was extracted, amended with biostimulants, and returned to the subsurface without aboveground treatment. After 7 months of treatment, 90% reduction in BETX in groundwater was observed. The success of this project raises the possibility that aboveground treatment of extracted groundwater may not always be required for closed-loop, in situ bioremediation systems. This change in regulatory attitude could result in a substantial reduction in costs.

Sites Not Upgradient from Sensitive Areas. At those sites where sensitive downgradient receptors are not threatened (i.e., sites located in flat areas with minimal natural groundwater flux or areas at considerable distance from sensitive receptors), plume control may not be required, especially if a nonliquid delivery approach such as air sparging (Sections 2.5.1.5 and 3.5.1.2.3) was implemented. Because groundwater is not extracted, the costs associated with air sparging are usually less than those associated with closed-loop delivery systems. Nevertheless, liquid delivery without groundwater capture (if permitted) or closed-loop treatment might be advantageous for flat sites with minimal natural groundwater flux as a means to enhance the rate of aquifer restoration by maximizing mass transport and mixing in the contaminated zone.

3.3.3.2 Geological Aspects

The physical aspects of the aquifer represent the environmental conditions under which the in situ treatment system must operate. If available, information on regional geology should be reviewed to gather insight into potential subsurface conditions. Also, if available, boring and monitoring well logs from investigatory activities at the site or nearby should be reviewed for information concerning grain-size distributions of aquifer sediments, stratigraphic variability, mineral composition, and the presence of bedrock.

If new borings are made as part of the remedial investigation, sediment samples from the aquifer should be collected from key aquifer strata for grain size and mineralogical analyses. These sediment samples will also be suitable for geochemical (Section 3.3.3.4) and biogeochemical (Section 3.3.3.5) analyses. Geological factors must be evaluated and taken into consideration to install a bioremedial system that will have a reasonable probability of success under site-specific conditions.

Grain-Size Distribution. The grain-size distribution of aquifer sediments influences pore-space volume and distribution and consequently, the aggregate permeability of the aquifer sediments. Basically, the finer the grain sizes, the more difficult it will be to achieve biostimulation. Problems have been reported for aquifer sediments with permeabilities less than or equal to 3.3×10^{-3} cm/sec (Lee et al., 1987). Furthermore, the in situ bioremediation program itself can reduce aquifer permeability. When treating fine-grained sediments, means to evaluate changes in aquifer permeability should be included (i.e. injection pressure changes, groundwater level fluctuation).

The grain-size distribution of the aquifer sediments should be determined or a preliminary characterization of sediment grain size may be conducted by inspecting existing boring or monitoring well logs.

If the results of geological and hydrological (see section 3.3.3.3) tests indicate adequate permeability, the data can be used in commercially-available aquifer-response models to evaluate various configurations for injection/extraction wells or trenches to provide a basis for remedial design (Piotrowski et al. 1993). Injection wells have been successfully used in predominately sand and gravel aquifers to biogeochemically influence extended regions of aquifers (e.g., > 200 feet in the downgradient direction over depths ranging to greater than 80 feet below ground surface) (Piotrowski, 1991; Piotrowski et al. 1994).

In low-permeability aquifers, trench systems may be necessary to extract and deliver groundwater at appreciable rates and across extended aquifer regions. While the rate of groundwater flux across a treatment zone can be modified to some extent in tight aquifers, its maximum rate (and the rate of biorestoration) will ultimately be constrained by the zone's aggregate permeability. Several vendors offer biodegradable slurries that allow ready installation of infiltration/extraction trenches and associated conveyance piping below the water table (Piotrowski et al. 1993). The costs associated with trench installation increase substantially with depth. At some sites, cost may favor injection/extraction wells over trenches even though trenches may produce a more rapid rate of biorestoration across extended aquifer regions.

Stratigraphic Variability. Stratigraphic variability can strongly influence the performance of an in situ bioremediation system. For example, it can result in channeling of groundwater through permeable zones such that the amendments used for in situ biostimulation may bypass the less permeable, yet contaminated zones. Channeling may especially occur in alluvial aquifers which can be comprised of lenses of markedly different permeabilities. The challenge at these sites is to devise a system that will deliver amendments to the most contaminated zones without substantial loss to permeable regions.

Although fine-grained lenses may adsorb organic contaminants and inhibit groundwater penetration, the fine-grained materials also tend to inhibit contaminant penetration into the lens. Consequently, a fine-grained lens may contain a veneer of contamination that may be treatable by passing groundwater amended with biostimulants over its surfaces. Promoting biodegradation of the veneer of contamination may render the lens an inconsequential source of groundwater pollution.

Mineral Composition of Sediments. The nature of the mineral composition of the aquifer sediments can influence both the physical and chemical aspects of a bioremedial design. Chemical aspects of the mineral content of aquifer sediments are discussed in Section 3.3.3.4. If sediment samples consist of more than 30 percent clay, it may be advisable to have portions of the samples analyzed for clay content using X-ray diffraction. The 2:1-layered silicate clays (e.g. illite, smectite, montmorillonite) tend to allow penetration of hydrophobic organic compounds into the clay-lattice structure which can interfere with microbial access to the compounds (for a review see pages 118-122 in Alexander 1994). In contrast, 1:1-layered silicate clays (e.g. kaolinite) do not allow penetration of organic contaminants.

If 2:1-layered clays are common, application of a surfactant may be required to enhance bioavailability of hydrophobic organic compounds (Leavitt et al. 1992; Bonin et al. 1994). However, the surfactant should be nontoxic, biodegradable, and not inhibit contaminant biodegradation. Moreover, hydraulic control of the contaminant plume may be required to prevent contaminant migration resulting from an enhancement of its mobility by surfactant application.

The Presence Of Bedrock. Those aquifers that contain bedrock pose unique challenges to bioremedial design. Bedrock fractures represent groundwater conduits for contaminant transport whose pathways are often difficult to map or predict. Improper selection of the sites of the injection/extraction components can easily render an in situ bioremediation system ineffective. A case study of bioremediation of bedorck was published by Bell and Hoffman (1991).

One method used to address this design complication is to apply a disruptive force to artificially increase the bedrock fractures, which increases aquifer permeability for enhanced biorestoration. Attempts have been made to increase permeability using hydraulic (Davis-Hoover et al. 1991; Vesper et al. 1994) and pneumatic (Schuring 1993) fracturing. However, care must be taken that the fracturing does not enhance contaminant mobility.

3.3.3.3 Hydrogeological Aspects

The ability to move groundwater through a contaminated zone can be vital to the success of an in situ bioremediation system. Information concerning groundwater flow can be obtained by conducting standard hydrogeological tests (e.g. draw-down tests, slug tests). These tests are often used to develop pump-and-treat remedial systems. Often overlooked during the design of a closed-loop in situ bioremediation system is the ability of the contaminant source area to accept the treated and amended groundwater. This information can be obtained by conducting standard percolation tests in the source area (Piotrowski et al. 1993).

As mentioned in Section 3.3.3.1, the information from draw-down and percolation tests can be used in commercially-available groundwater-response models to evaluate various extraction and delivery configurations to provide a basis for a closed-loop design. As discussed further in Section 3.3.3.4, a draw-down test also provides the opportunity to collect information on groundwater chemistry and contaminant concentration that can be important to the design of an aboveground water treatment system.

Finally, air sparging can produce localized mounding in the surface of the water table in the vicinity of the sparging well. The mounding represents a groundwater gradient that can increase groundwater flow and could elicit regulatory concern that enhanced contaminant migration may result. Therefore, groundwater level measurements during a pilot test of air sparging may be useful to evaluate this possibility and provide an indication if plume control may be warranted.

3.3.3.4 Geochemical Aspects

The geochemistry of the aquifer sediments are also important for bioremedial design. Topics discussed in this section include sediment contaminant concentrations, considerations of iron-rich aquifers, evaluations of sediment redox capacities, and chemical analyses during extended aquifer draw-down tests.

Sediment Contaminant Concentrations. Of utmost importance to regulatory and remedial design is the concentration of adsorbed contamination. The adsorbed contaminants (which can exist as organic coatings, microglobules, or free product) represent long-term sources of groundwater contamination. If the adsorbed contaminants are not treated along with the groundwater, extended treatment times will be required. Therefore, an evaluation of the adsorbed contaminant concentrations in and around the contaminant source area should be performed during the remedial investigation.

For organic liquids with specific gravities < 1.0 g/cc (light nonaqueous phase liquid, LNAPL), typically the upper strata of the aquifer will be impacted. For organic liquids with specific gravities > 1.0 g/cc (dense nonaqueous phase liquid, DNAPL), extended vertical sections of aquifer strata may be impacted, especially in the vicinity of the contaminant source area. During the advancement of soil or well borings, it is advisable to collect samples from those strata likely to be most contaminated and analyze for contaminants. When analyses of contaminant concentration and nutrients (e.g. Kjeldahl nitrogen, phosphorus, total organic carbon) are coupled, potential nutrient requirements for in situ contaminant biodegradation can be assessed.

Considerations for Iron-Rich Aquifers. Iron-rich sediments under reducing conditions can generate groundwater enriched with dissolved ferrous iron. The presence of dissolved iron can affect the operation of both above- and below-ground components of an in situ bioremediation system. If an oxidative approach is used, conversion of the soluble ferrous to the insoluble ferric species can result in excessive formation of iron precipitates and flocs that can clog pore spaces in aboveground filters, pump-intake screens, and between aquifer sediment particles.

Iron precipitation interferes with groundwater flow. The operation plan for the bioremedial system at such sites may require frequent changes of abovegound filters and in situ pump screens and the frequent application of chemical or physical means to break-up or dislodge iron flocs in the formation. Although iron precipitation may be an operational concern in the early stages of oxic treatment of an iron-rich, contaminated aquifer, once oxic conditions develop throughout the contaminated zone, iron will tend to precipitate and remain in the aquifer. Consequently, iron precipitation problems in the aboveground components will abate.

Finally, the presence of iron can influence the selection of the form of phosphorus fertilizer used in phosphorus-deficient aquifers. Brown and Norris (1994) suggest the use of tripolyphosphate to reduce the potential for iron precipitation.

Sediment Redox Capacities. The oxidation-reduction (redox) capacities of aquifer sediments may resist efforts to adjust redox conditions in the aquifer (Barcelona and Holm 1991). Appreciable quantities of an oxidant may be consumed by abiotic oxidations of the reduced species before measurable changes in groundwater redox are observed. This possibility must be taken into consideration during the selection of the oxidant for the in situ bioremedial system.

If anticipated or known that the aquifer sediments will consume a large amount of the oxidant in abiotic redox reactions, it may be advisable to initially use high concentrations of the oxidant, hydrogen peroxide, to quickly satisfy the abiotic oxygen demand. Once the abiotic demand has been met, lower peroxide concentrations or alternate oxygen sources may be used to reduce operational costs. Although high concentrations of hydrogen peroxide may sterilize regions of the aquifer close to the injection point, recolonization probably occurs after peroxide addition is terminated. Additional discussion of means to oxidize aquifers is presented in Section 3.3.3.5.

Chemical Analyses During Extended Aquifer Draw-Down Tests. As mentioned in Section 3.3.3.3, important information on groundwater chemistry and contaminant concentration may be obtained during an extended aquifer draw-down test. Many times the aboveground treatment system is sized based on the results of point-in-time measurements of groundwater contaminants. Because the concentration of organic contaminant in groundwater declines as more and more water is extracted, aboveground systems based on samples collected during the early phase of pumping may be oversized. To avoid sizing problems, groundwater samples should be collected at intervals during the course of an extended draw-down test (e.g at 12 and 24 hours during a 24-hour test) and analyzed for contaminant concentration (Piotrowski et al. 1993).

In addition to contaminant concentration, the groundwater samples should be analyzed for dissolved solids, alkalinity, hardness, anions, cations, and total Kjeldahl nitrogen. Hard, alkaline water with a high concentration of dissolved solids may cause scaling problems in system components. If so, design measures can be implemented to reduce the scaling problem.

The results of anion, cation, and total Kjeldahl nitrogen analyses can be used to assess ionic balance and nutrient relationships (iron, ammonia, and potassium concentrations can be measured in the cation scan; nitrate, nitrite, and phosphate can be measured in the anion scan). The results of the analyses in conjunction with sediment analyses may indicate that some form of nutrient addition may benefit the in situ bioremedial process.

3.3.3.5 Biogeochemical Analyses

The results of biogeochemical analyses of groundwater and aquifer sediments can be used to assess the potential for successful bioremediation. When oxygen is used as the terminal electron acceptor, the means by which oxygen is delivered will depend on the redox of the aquifer.

Groundwater Analyses. In addition to the chemical analyses discussed in Section 3.3.3.4, groundwater may be analyzed for dissolved oxygen concentrations and other indicators of microbial activity. Because microorganisms are usually attached to surfaces rather than free in the aqueous phase, the most accurate assessment of microbial numbers and biodegradation potential should be determined using samples of subsurface material.

Oxygen concentrations can be monitored in wells located upgradient from (uncontaminated) and within the plume. If dissolved oxygen is measured in a transect of monitoring wells extending from upgradient areas into the plume, a typical pattern is observed. Upgradient of the plume, dissolved oxygen is detectable (2-8 mg/L), whereas within the plume dissolved oxygen concentrations are < 1mg/L. The absence of dissolved oxygen along the transect is a result of aerobic biodegradation of contaminants (Piotrowski 1989; Piotrowski 1991). Although simple, this method has been used successfully at a Superfund site to convince EPA to allow a pilot study of in situ bioremediation (Piotrowski 1989).

Other indicators of microbial activity that can be measured in groundwater include the concentrations of ferrous iron, methane, and electron acceptors other than oxygen (see Section 3.2.2.3, Estimates of Biodegradation Using the Expression of Electron Acceptor Demand).

Sediment Analyses. Samples for microbial analyses should be incubated at the in situ temperature of the groundwater of the site. Various enumeration and activity measurements are available, such as total counts, viable counts, counts of contaminant-specific degraders, and radiolabelled biodegradation studies. A broad range of methods has been summarized (Kemp et al. 1993). Caution should be exercised in extrapolating the results of laboratory experiments to the field. The results of in situ pilot studies are the best indicators of remedial time requirements.

Means to Deliver Oxygen. Because of the redox capacities of sediments (Section 3.3.3.4), the means used to meet chemical and biological oxygen demands of the contaminated subsurface will influence the remedial design and costs.

Hydrogen peroxide has been used successfully to supply oxygen to contaminated aquifers. Although relatively expensive, it may be especially useful for quickly eliminating abiotic oxygen demand in reducing aquifers. Also, ozone has been used. An added benefit of these oxidants is that they can also chemically oxidized organic contaminants.

Direct oxygen supply with air, oxygen generators, or liquid oxygen have been used successfully. The oxygen is delivered to the groundwater by sparging aboveground or in situ (Brown and Jasiulewicz 1992). Unfortunately, these forms of oxygen addition are not very efficient. An alternative approach, in situ delivery of oxygen microbubbles (colloidal gas aphrons), holds promise as a more efficient means to deliver oxygen (Michaelson and Lofti, 1990).

Finally, another relatively new approach involves using semi-permeable tubules which are charged internally with oxygen gas. As groundwater flows over the surface of the tubules, oxygen molecules diffuse across the membranes and directly enter the groundwater without bubble formation (Semmens et al. 1991). These systems are 100 percent efficient in oxygen delivery. Field tests of aboveground systems indicate that dissolved oxygen concentrations of >40 mg/L can be created, even at water flow rates of 250 gallons per minute (Piotrowski et al. 1994).

In summary, the extent of reducing conditions in the contaminated aquifer will be the driving force in the selection of the means to deliver oxygen. At some sites, it may be advisable to use a series of oxygen sources (i.e., hydrogen peroxide or ozone followed by air or oxygen sparging) over the course of remediation to convert the aquifer to oxic conditions and stimulate elevated rates of aerobic biodegradation of the organic contaminants.

3.4 *Natural Bioattenuation of Hazardous Organic Compounds in the Subsurface*

3.4.1 Patterns of Natural Bioattenuation

Occasionally organic contaminants enter the subsurface as spills of liquids miscible in water, or as true solutions in water. Examples include ethylene glycol, used as a de-icing agent for aircraft, and organic solvents containing oxygen, such as acetone or butanol, used in the pharmaceutical industry. Such releases tend to move with the flow of groundwater and be flushed away from the point of release.

Most organic contaminants, however, enter the subsurface as an oily liquid, such as a fuel spill or a release of a chlorinated solvent. Groundwater moving through the material dissolves a small portion of the contaminant, which becomes a plume of groundwater contamination. Because the contaminant mass in the oily material is much greater than that dissolved in the groundwater, the spill can continue to maintain the plume more or less indefinitely. As the plume moves away from its source, natural biological processes may attenuate the contamination in the groundwater.

There are three patterns of natural biotransformation of hazardous organic compounds in the subsurface. A wide variety of organic materials are degraded easily with oxygen or nitrate as the electron acceptor. Under aerobic and dentrifying conditions, microbial populations quickly adapt and reach high densities. These conditions result in the first pattern, in which the rate of biodegradation quickly becomes limited by the rate of supply of some nutrient, not by the microbial capacity to degrade the contaminant.

Some organic contaminants can also be degraded in the absence of oxygen. In the absence of oxygen, their degradation follows a second pattern; the rate of degradation usually is limited by the reaction rate of the active microorganisms. Reaction rate is related to substrate concentration by a hyperbolic function, and is best described by Monod or Michaelis-Menton kinetics. Biodegradation under methanogenic, sulfate-reducing, and iron-reducing conditions follow this pattern. Supplies of carbonate and iron minerals usually are not limiting, but if the supply of sulfate is depleted, then biodegradation under sulfate-reducing conditions would start to follow the first pattern.

In the third pattern, the organic compound serves as an electron acceptor, instead of electron donor or carbon source, its common role. Under this pattern, the rate of transformation of the organic compound is controlled by the electron acceptor demand exerted by other organic compounds and by competition from other, more conventional electron acceptors.

3.4.2 Aerobic Biotransformation of Easily Degraded Compounds

Hadley and Armstrong (1991) compared the number of water supply wells in California that were contaminated with benzene, which is easily degraded in aerobic groundwater, with the number contaminated with trichloroethylene or tetrachloroethylene, which are not easily degraded aerobically. Because of the ubiquity of underground storage tanks that leak gasoline, one might expect that more wells would be contaminated with benzene than with the chlorinated solvents; however, the solvents were encountered more frequently, and at higher concentrations (table 3.5). This finding suggests that natural biodegradation was removing the benzene from aerobic groundwaters in California.

Wilson et al. (1985) used microcosms to estimate the potential for biotransformation of naphthalene, methylnaphthalenes, dibenzofuran, and fluorene in a plume of contaminated groundwater originating from a disposal lagoon for wood-preserving wastes. The contaminants were not biodegraded in uncontaminated aquifer material from the site; however, rapid

Table 3.5
Relative Occurrence of Benzene, Trichloroethylene, and Tetrachloroethylene in Water Supply Wells in California

Compound	Number of Wells	Median Concentration	Range of Concentrations
		-------(μg/L)------	
Benzene	9	0.2	0.1 - 1.1
Trichloroethylene	188	3.2	0.1 - 538
Tetrachloroethylene	199	1.9	0.1 - 166

Hadley and Armstrong 1991

degradation was detected in aerobic aquifer material collected from the margin of the plume. First-order rate constants were on the order of 1.0% per week. The plume had persisted for many years, which was inconsistent with the rapid kinetics of biodegradation in oxygenated groundwater. An examination of the geochemistry of the groundwater revealed that the plume was persistent in anaerobic water, but the organic contaminants were entirely depleted in water from the plume that had been oxygenated by admixing with uncontaminated groundwater.

As a consequence of microbial adaptation in the aquifer, the rate of biodegradation of organic contaminants was not limited by the metabolic activity of the microorganisms. The rate of biodegradation was limited by the rate at which oxygen was supplied to the plume. The pH was low (<4), and there was no evidence of anaerobic biodegradation.

Borden et al. (1989) confirmed the laboratory microcosm observations by conducting injection and withdrawal tests in the plume of creosote contamination. There was minimal degradation of the PAHs when the water injected into the aquifer did not contain oxygen; however, rapid and extensive degradation of these compounds was detected when oxygenated groundwater was injected. Oxygen was not limiting for biodegradation at concentrations as low as 0.7 mg/L. The injection and withdrawal tests were conducted over a period of a few days. On this time scale, the minimum concentration of total PAHs that could be achieved was 30 to 70 μg/L. But, the same compounds were undetectable in monitoring wells in the naturally-bioremediated portion of the plume. Presumably, much lower concentrations can be achieved with a longer residence time.

Mixing processes in aquifers that can blend oxygen into a plume include dispersion and diffusion. Dispersion is proportional to advective flow in the aquifer; the extent of mixing is proportional to the distance the plume moves. Diffusion is proportional to residence time of the plume in the aquifer. Borden and Bedient (1986) constructed a mathematical model, BIOPLUME, of aerobic bioremediation in aquifer that assumed an adapted microbial population was present, that oxygen was required for biodegradation, and that oxygen transport was the rate-limiting step. The rate of reaeration is dependent on the saturated thickness of the aquifer containing hydrocarbon contamination and the vertical dispersion coefficient. These parameters were used to estimate a site-specific pseudo first-order reaeration constant, which predicted an apparent first-order biodegradation rate constant of 25% per year.

The model adequately simulated the plume of contamination from the creosote waste lagoon, which was forty years old (Borden et al. 1986). The plume had significantly attenuated after traveling 400 m (437 yd), requiring approximately twenty years. Simulations suggested that the plume had already reached its maximum extent.

Chiang et al. (1989) used BIOPLUME II, a commercial version of the model available from Rice University in Houston, Texas, to evaluate the natural degradative characteristics of a plume of alkylbenzenes originating from the flare pit of a natural gas plant. Microbes in the aquifer had adapted to degrade the contaminants. As a consequence of adaptation, oxygen was absent at locations where alkylbenzenes were present. At locations where oxygen was present, alkylbenzenes were absent. BIOPLUME II adequately predicted the concentrations of alkylbenzenes in monitoring wells at the site. The apparent first-order biodegradation rate constant was 35% per year.

Barker, Patrick, and Major (1987) created a plume of alkylbenzenes in a sandy water-table aquifer in Canada, then monitored the attenuation of the plume as it moved through the natural gradient. The uncontaminated water in the aquifer was oxygenated. The plume contained benzene, toluene, o-xylene, and chloride as a conservative tracer. The plume initially contained a concentration of 7.6 mg/L alkylbenzenes in a volume of 1,800 L. The initial plume was approximately 3 m (9.8 ft) long in the direction of ground-water flow. The plume attenuated in 1.2 years after moving 30 m (98.4 ft) through the aquifer.

The alkylbenzenes were removed at the same rate without a lag period. Initial removals were apparently zero-order. As the plume segregated into regions with higher and lower hydraulic conductivity, the rate of removal of the suite of alkylbenzenes became pseudo first-order. There was good correlation between the field data and laboratory microcosm studies, but the authors concluded that it was fortuitous.

Klecka et al. (1990) described a plume of phenolic compounds and PAHs originating from buried wastes from a charcoal manufacturing plant. The aquifer was aerobic, except for the region that contained the plume of contamination. The plume attenuated within 60 to 100 m (66 to 109 yd) of its source. Groundwater would require 0.7 to 1.4 years to move 100 m (109 yd). A one-dimensional model, called BIO1D, that predicts biodegradation rates along a flow path in an aquifer was used to estimate the relative importance of sorption and biodegradation for contaminant removal.

The contribution of sorption to contaminant removal was determined to be negligible through a sensitivity analysis performed with laboratory sorption data. Microcosm studies, however, showed that microbes in the aquifer were capable of removing the organic contaminants at rates that would explain the attenuation of the plume. Although the reaction kinetics in the microcosms corresponded to the rate of attenuation in the field, the authors concluded that the "biological reaction rates [in the plume] are controlled by the availability of dissolved oxygen in the zone of contamination."

3.4.3 Anaerobic Biodegradation of Contaminants as Carbon or Energy Sources

It is now well established that aromatic organic compounds, such as the alkylbenzenes, certain simple PAHs, and some nitrogen-containing heterocyclic organic compounds, can be degraded in groundwater in the absence of oxygen (Grbic-Galic 1990). The aromatic compounds are oxidized first to phenols or organic acids, then transformed to long-chain volatile fatty acids, which are finally metabolized to methane and carbon dioxide. Adaptation of the microorganisms to degrade the contaminants is slow, requiring months to years. Destruction of the hazardous contaminants, such as the alkylbenzenes, is associated with accumulation of fatty acids, production of methane, solubilization of iron, and reduction of nitrate and sulfate (Cozzarelli, Eganhouse, and Baedecker 1990; Wilson et al. 1990).

When oxygen is absent, then nitrate, sulfate, carbonate, and iron III can serve as terminal electron acceptors (see Subsections 3.2.1 and 3.2.2). Plumes of aromatic compounds do not contain nitrate (John Wilson, personal experience; see B. H. Wilson et al. 1990 for an illustration), indicating that microbial adaptation to use nitrate as a terminal electron acceptor occurs readily, and ambient concentrations of nitrate are quickly consumed. Natural attenuation of plumes of aromatic contaminants through nitrate respiration should be similar to oxidative biodegradation.

Water table aquifers, particularly in agricultural areas, contain considerable electron-accepting capacity in the form of nitrate. A typical water table aquifer may contain 5 mg/L of dissolved oxygen and 5 mg/L nitrate-nitrogen. There is, however, potentially three times the electron-accepting capacity in the nitrate as in oxygen. To the authors' knowledge, the potential contribution of nitrate has not been factored into models of the natural aerobic attenuation of organic contaminants.

3.4.3.1 Sulfate-Reducing Conditions

Plumes of aromatic compounds commonly contain sulfate, and, depending on its concentration, sulfate can also be an important electron acceptor. Acton and Barker (1992) monitored the attenuation of benzene, toluene, o-xylene, m-xylene, ethylbenzene, and 1,2,4- trimethylbenzene in a forced-gradient injection experiment in a plume from an active landfill owned by the City of North Bay, Ontario, Canada. Leachate was collected from the plume, amended with approximately 200 µg/L of the six organic compounds, bromide as a tracer, and 12 mg/L sulfate, and then re-injected into the aquifer. After a 30-day lag period, toluene and m-xylene were degraded, but the other organic compounds were not.

Thierrin et al. (1992a, b) described the bioattenuation of a plume from a gasoline spill in Perth, Australia, under sulfate-reducing conditions. A plume of benzene in groundwater extended at least 420 m (459 yd) from the spill, while toluene was completely degraded within 200 m (219 yd). Uncontaminated groundwater in the aquifer contained more than 20 mg/L of sulfate. The concentration of sulfate in groundwater from the plume was below that detected in uncontaminated groundwater from the site (Thierrin et al. 1992b). In an experiment, 400 L of groundwater was extracted from the plume and amended with bromide and a solution of fully deuterated benzene, toluene, p-xylene, and naphthalene as tracers of contaminant transport and biodegradation; the amended groundwater then was injected back into the plume 80 m (87 yd) from the source area. The deuterated compounds could be distinguished from the natural aromatic compounds with good precision and sensitivity using mass spectrometry.

The concentration of the bromide and deuterated aromatic compounds was monitored over time in a series of cluster wells that sampled water above, within, and below the plume (table 3.6 on page 3.53). The data are from a series of vertically-stacked cluster wells down gradient of the spill area.

Table 3.6
Rates of Bioattenuation Under Sulfate-Reducing Conditions in a Plume From a Gasoline Spill

Depth Below Water Table / Travel Time of Bromide	Benzene	Toluene	*p*-Xylene	Naphthalene
meters/weeks	---------percent removed per week---------			
Slightly oxygenated water at the upper edge of the contaminant plume				
0.75/>10	above plume	above plume	above plume	above plume
1.0/6.2	6.5	>81	>32	>32
Anoxic/Sulfate-reducing Zone				
1.25/6.5	3.0	35	1.7	>40
1.5/5.8	1.2	6.6	none	6.6
1.75/5.4	none	7.0	3.2	14
2.0/5.5	none	5.1	2.1	15
2.25/5.8	none	none	1.6	12
2.5/5.9	1.4	3.6	2.5	18
2.75/5.9	20	8.3	?	30
3.0/>10	below plume	below plume	below plume	below plume

Thierrin et al. 1992b

Total attenuation was corrected for sorption and dilution to estimate bioattenuation. Rates of removal were greatest at the margins, perhaps because the supply of sulfate and oxygen there was greater. In the interior of the plume, the attenuation of benzene was actually slightly less than that of bromide.

Thierrin et al. (1992a) compared the bioattenuation of the tracer plume of deuterated aromatic compounds to the attenuation required to fit a model of the full-scale plume. These two independent assessments of the behavior of aromatic compounds were in close agreement for benzene and toluene (table 3.7 on page 3.54).

Benzene was not removed in the heart of the plume. The rates of bioattenuation of toluene, the xylenes, a trimethylbenzene, and naphthalene were adequate to degrade these compounds within a few hundred meters of travel (table 3.7 on page 3.54).

Table 3.7

Comparison of Kinetics of Bioattenuation of an Artificial Plume of Deuterated Aromatic Organic Compounds and a Full-Scale Plume Under Sulfate-Reducing Conditions

Organic compound	Tracer test with deuterated organics in the field	Required to make computer model fit field data
	---percent per week depleted---	
Benzene	<0.6	<0.6
Toluene	4.8	4.0
Ethylbenzene	no data	2.1
o-Xylene	no data	3.9
p-Xylene	2.2	not resolved
m+p-Xylene	no data	2.9
1,3,5-Trimethylbenzene	no data	2.7
Naphthalene	15	3.0

Thierrin et al. 1992a

3.4.3.2 Methanogenic Conditions

The important electron sinks in many anaerobic plumes are the iron III and mixed valence iron minerals in the aquifer matrix (Lovley 1991) and bicarbonate in carbonate/bicarbonate-buffered groundwaters. In the case of carbonate and iron III, the supply of electron acceptor is not limiting for contaminant biodegradation. The metabolic activity of the microorganisms becomes the rate-limiting step. As a consequence, the rate of reaction is controlled by the density of active organisms and by the concentration of metabolizable compounds.

Godsy and his associates in the U.S. Geological Survey have studied methanogenic transformation of a series of phenols in a plume originating from a disposal lagoon for wood-creosoting wastes. Microcosm studies were used to estimate the kinetic constants of degradation of four phenols (Godsy, Goerlitz, and Grbic-Galic 1992b).

The concentrations of phenols at which microbial growth is one-half maximum (K_s) were low, near 1 mg/L (table 3.8 on page 3.55). The maximum growth rate (μ) was slow in comparison with that of other microorganisms. The yield coefficients (transformation of substrate to biomass, Y)

were low, and the rate of self-consumption (K_d) was low or undetectable. Further, there seemed to be an upper limit to the microbial biomass in the microcosms (Godsy, Goerlitz, and Grbic-Galic 1992b). Although Monod growth kinetics predicted that biomass would reach much higher concentrations, the final biomass in the microcosms ranged from 0.030 to 0.044 mg/L. Apparently, there is some density-dependent constraint, such as predators of the phenol-degrading microorganisms, that limits the development of biomass.

Table 3.8

Kinetics of Biodegradation of Organic Contaminants in Microcosms Under Methanogenic Conditions

Compound	μmax (per day)	Ks (mg/L)	Y (mg biomass/ mg phenol)	Kd (per day)
Phenol	0.11	1.3	0.004	0.001
2-Methylphenol	0.044	0.25	0.003	0.002
3-Methylphenol	0.10	0.55	0.002	0.000
4-Methylphenol	0.10	3.3	0.042	0.000

Adapted from Godsy et al. 1992b

These observations support a provisional paradigm for the function of anaerobic biodegradation in aquifers in which the concentration of active biomass is controlled by some density-dependent constraint. Under this scenerio, the density of active organisms is uniform throughout the plume and is independent of the concentration of aromatic contaminants. The degradation of an individual aromatic contaminant will be zero-order at concentrations above 1 mg/L and first-order at lower concentrations. Therefore, it should be possible to model bioattenuation of aromatic contaminants with a simple one-dimensional model, such as BIO1D, calibrated from microcosm studies or from field estimates of contaminant attenuation.

Godsy, Goerlitz, and Grbic-Galic (1992a) found a high degree of correlation between the behavior of laboratory microcosms and the field-scale plume of creosote contamination mentioned earlier (table 3.9). The time taken for water to move a certain distance from the source along a flow path was compared to the residence time in the microcosm.

Table 3.9

Comparison of Bioattenuation of Organic Contaminants in a Plume of Contaminated Groundwater and in Microcosms Simulating the Plume Under Methanogenic Conditions

Behavior	Residence in Microcosm (days)	Residence in Plume (m along flow path)
C_3 to C_6 volatile fatty acid converted to acetate and methane.		
Benzoic acid degraded	0 to 50	0 to 50
Phenol degraded	50 to 99	53 to 98
Quinoline and isoquinoline degraded	0 to 50	53 to 98
2-,3-,4-Methylphenol degraded	100 to 200	53 to 125
Quinoline and isoquinoline degraded	100 to 180	produced

The average seepage velocity in the plume was 1.0 m per day

Plumes of alkylbenzenes from fuel spills behave the same way. Wilson et al. (1990) described the methanogenic bioattenuation of benzene, toluene, and xylenes (BTX) in groundwater contaminated by a spill of aviation gasoline. Attenuation of total BTX was measured quarterly over a 4-year period in wells along a flow path running through the centerline of the plume. Removal followed first-order kinetics, varying from 10 to 34% per week. When a purge well field went on line, one of the monitoring wells was isolated from the source area. Alkylbenzene concentrations dropped rapidly in the isolated well. Interestingly, toluene disappeared more than twice as rapidly as benzene. Shortly after the BTX compounds disappeared in the isolated well, core material was acquired from that region for a laboratory microcosm study. The kinetics of bioattenuation were very similar for in situ measurements along the flow path, in stagnant water near the isolated monitoring well, and in the microcosm study (table 3.10).

The independent field-scale estimates of bioattenuation were in agreement within a factor of three, and those of the microcosm, within a factor of ten. The seepage velocity was known within a factor of two.

Cozzarelli, Eganhouse, and Baedecker (1990) reported anaerobic bioattenuation of alkylbenzenes in a plume of contamination from a spill of crude oil in Minnesota. Methane was detected in the groundwater and long-chain volatile fatty acids accumulated. Bioattenuation in monitoring wells in a transect along the centerline of the plume was reported. Toluene and o-xylene disappeared without a lag period, indicating that the microorganisms had adapted to degrade the compounds. Both were depleted within 12 m (40 ft) of the source area. Ethylbenzene degradation began after toluene

Table 3.10

Bioattenuation of Benzene, Toluene, and Xylenes in Microcosms and Methanogenic Groundwater at an Aviation Gasoline Spill Site

Compound	Microcosms	Flow path	Isolated well
	-----percent per week-----		
Benzene	50	5	17
Toluene	30	130	47
m+p-Xylene	40	---	---
o-Xylene	50	---	---
All Xylenes	---	3	10

and o-xylene had disappeared. Benzene was depleted without a lag period in the presence of toluene and o-xylene; however, the rate of benzene depletion slowed greatly after these compounds disappeared. The average seepage velocity in the plume is 0.1 m/day (0.3 ft/day) (personal communication, Philip Bennett, U. of Texas at Austin). This value was used to express the attenuation between wells as a first-order rate constant (table 3.11 on page 3.58).

When the plume mixed with oxygenated groundwater, elevating the dissolved oxygen concentration above 1.0 mg/L, all the aromatic hydrocarbons disappeared. This occurred within 100 m (109 yd) of the spill.

Wilson, Kampbell, and Armstrong (1993) studied bioattenuation of aromatic hydrocarbons in a plume produced from gasoline leaking from a underground storage tank. Methane was detected in the groundwater and acetate accumulated. Bioattenuation of alkylbenzenes was estimated between the spill area and monitoring wells located 100 days downgradient and 200 days downgradient. The rates of anaerobic bioattenuation were very similar to those rates in the crude oil spill (table 3.11) and the aviation gasoline spill (table 3.10 on page 3.57).

3.4.4 Biodegradation and Organic Contaminants as Electron Acceptors

In the absence of oxygen, halogenated organic compounds can serve as electron acceptors. The most important example of this process is the sequential reductive dehalogenation of tetrachloroethylene to TCE, to *cis* or *trans*-dichloroethylene, to vinyl chloride, and finally to ethylene, ethane, and methane (Vogel and McCarty 1985; de Bruin et al. 1992).

This process is well-documented qualitatively, but there have been few full-scale studies of the kinetics of dehalogenation. Researchers from the Trenton, New Jersey Office of the U.S. Geological Survey and the R.S. Kerr Laboratory of the U.S. EPA compared the kinetics of TCE dechlorination in microcosms to the rate of dechlorination in a large plume of TCE at the Picatinny Arsenal in New Jersey (Imbrigiotta et al. 1991).

Table 3.11

Rates of Bioattenuation of Alkylbenzenes in Plumes from a Gasoline Spill and a Crude Oil Spill Under Methanogenic Conditions

Compound	Crude Oil Spill	Gasoline Spill (0 to 100 days from source)	Gasoline Spill (100 to 200 days from source)
		-----percent per week-----	
Benzene	12	13	3
Toluene	50	21	26
o-Xylene	40	16	8
p-Xylene	--	15	9
m-Xylene	--	13	8
Ethylbenzene	19	12	13

Trichloroethylene was discharged from a plating shop to a shallow dry well from 1973 to 1978. A presumed source area of nonaqueous phase liquids produced a plume of TCE, *cis*-dichloroethylene, and vinyl chloride. The plume extended 488 m (1,600 ft) from the source area to the point of groundwater discharge to surface drainage. Although the average residence time for water was 0.7 to 1.5 years (Voronin 1991), the concentration of TCE in the plume did not decline significantly within 4.6 years after TCE discharge to the aquifer ceased (Imbrigiotta et al. 1991); this indicates that the groundwater plume was stable and sustained by dissolution of oily phase liquids in the source area.

The highest concentrations of TCE were in the water table aquifer, just above the confining layer at the mid-point between the source and point of discharge. The concentration of TCE declined an order of magnitude from the area of highest concentration to the point of discharge to surface water. First-order rate constants for TCE attenuation, estimated from average seepage velocities in the plume, ranged from 3 to 9% per week. Laboratory data from Wilson (1988), Wilson et al. (1991), and Ehlke et al. (1991) are presented in table 3.12 (on page 3.60). The rates of TCE disappearance in the microcosms varied, but, in general, were consistent with the rate of attenuation at field scale. On the other hand, *cis*-dichloroethylene (DCE) only disappeared in microcosms constructed from regions of the aquifer containing significant amounts of it.

A variety of organic compounds can serve as electron donors for biological reductive dechlorination (DiStefano, Gossett, and Zinder 1992). Some microcosms were amended with volatile fatty acids, toluene, and p-cresol in an attempt to stimulate reductive dechlorination (table 3.12 on page 3.60) (Wilson et al, 1991; Ehlke et al. 1991). These compounds actually suppressed reductive dechlorination in some experimental treatments. To date, the chemical reducing agent that facilitates biological reductive dechlorination in the TCE plume at Picatinny Arsenal has not been identified; however, the process resulted in a substantial destruction of contaminant mass before the TCE discharge to surface water.

Table 3.12

Depletion of Trichloroethylene (TCE) and cis-Dichloroethylene (DCE) in Microcosms Constructed With Core Material From the TCE Plume at Picatinny Arsenal, New Jersey

Incubation (Weeks)	TCE Alone	TCE with Amendments	DCE Alone	DCE with Amendments
	----(percent per week)----			
Near source area, coarse sand and fine gravel				
40	5.3	3.9		
14			<0.1	
Near mid point, the area of highest TCE and cis-DCE concentration, fine sands and silts.				
25	0.9	0.6		
14			18, <0.1	3.3
Near the point of discharge to surface drainage, fine sands and silts.				
40	5.8	11		
25	4.6			
14			6.0	
Uncontaminated material near point of discharge to surface water, medium sands.				
25	5.1			
14			0.5	

3.5 ***Bioremediation Processes***

The following physical processes can be used for the biological remediation of liquid wastes, sludges, and contaminated surface soils, subsurface sediments, and air:

- in situ treatment of contaminated soil or sediment; and
- ex situ treatment of contaminated material in an aboveground reactor or prepared bed. The remediated materials are disposed of at an appropriate site.

In situ processes include land treatment, bioventing, liquid delivery, and air sparging; their developmental status is shown in Table 3.13 (on page 3.61). Ex situ processes include treatment in aboveground reactors, land treatment, composting, soil piles, and biofilters; their developmental status is shown in Table 3.14 (on page 3.62).

The selection of the biological treatment process is based on the physical and hydrogeological characteristics of the site, the chemical nature of the waste, and the clean-up levels required. For example, in situ or ex situ land treatment or composting generally might be feasible for bioremediation of soil contaminated with diesel fuel; however, because of land constraints or the presence of a seasonally-high water table in a particular site, in situ land treatment may not be feasible. A number of summaries, reports, and texts describe the use of these processes in treating diverse hazardous and non-hazardous wastes and contaminated soils (American Petroleum Institute 1983; Brown, Evans, and Frentup 1983; Sims, Sims, and Matthews 1989; Overcash and Pal 1979; Loehr and Malina 1986; US EPA 1990; Thomas, Ward et al. 1992; Nyer 1992).

Soil, sediment, or a bulking material is used in each of these three processes to facilitate the bioremediation activity. Soil and sediment provide nutrients, microorganisms, and buffering capacity for in situ and ex situ land treatment systems, and bulking material increases porosity to facilitate oxygen transfer in composting. In many cases, the soil or sediment is also being bioremediated. Generally, a mixture of contaminants is treated. The complex bonding forces exhibited by various soil fractions, particularly

Table 3.13
Developmental Status of In Situ Bioremediation Technologies

Technology	Waste Treated	Developmental Status		
		Laboratory/ Developing	Demonstrated in Field Trial	Proven Full Scale
Natural bioattenuation	Petroleum hydrocarbons		+	
	Chlorinated solvents			
Land treatment	Petroleum hydrocarbons			+
	PAHs; sludges; contaminated soils			
Bioventing	Petroleum hydrocarbons			+
	Chlorinated solvents	+		
Liquid delivery (aerobic)	Petroleum hydrocarbons			+
	Nonchlorinated solvents			+
	Chlorinated solvents		+	
	Pesticides	+		
Liquid delivery (anaerobic)	Monoaromatic hydrocarbons		+	
	Chlorinated solvents; PAHs	+		
Air sparging	Petroleum hydrocarbons		+	
	Chlorinated solvents	+		

Table 3.14
Developmental Status of Ex Situ Bioremediation Technologies

Technology	Waste Treated	Developmental Status: Laboratory/ Developing	Demonstrated in Pilot Study	Proven Full Scale
Suspended growth	Aromatic and aliphatic hydrocarbons; mono- and Dichlorinated aromatic and aliphatic hydrocarbons (>50 mg/L)			+
Fixed-film	Petroleum hydrocarbons			+
Submerged fixed-film	Aromatic and aliphatic hydrocarbons			+
Activated carbon-based	Pesticides; volatile organic compounds; low Concentrations of chlorinated organic compounds			+
Slurry-phase	PAHs; chlorinated solvents; munitions; pesticides		+	+
Land treatment	Petroleum hydrocarbons; PAHs			+
Soil-pile	Petroleum hydrocarbons; PAHs; nonchlorinated Solvents; chlorobenzenes		+	+
Composting	Munitions		+	
Biofilter	Hydrocarbons; chlorinated alkanes, alkenes	+		+

clays and organic matter, can affect the treatability of organic compounds in soils (see Subsection 3.2.1.2.5 Contaminant Bioavailability and Section 3.3 Site Characterization Relevant to In Situ Bioremediation). Therefore, an understanding of soil characteristics is important to the effective application of these processes.

Other parameters that affect the performance of bioremediation systems include the following:

- *Equalization.* For ex situ bioremediation processes, equalization minimizes variations in the contaminant load. This is important because biological treatment is sensitive to variations in organic loadings.
- *Nutrient Management.* Nutrient management is important because an insufficient amount of nutrients will slow removal of organic compounds. The principal inorganic nutrients are nitrogen and phosphorus. Trace amounts of potassium, calcium,

sulfur, magnesium, iron, and manganese are also required for optimum biological growth.

- *Oxygen Supply and Aeration.* An adequate supply of oxygen is critical to an environment in which aerobic organisms can grow and metabolize the organic material. The oxygen can be provided as atmospheric oxygen or in the form of oxygen-supplying compounds (e.g., peroxides). In composting, diffusion of oxygen is related to the physical and chemical characteristics of the compost. If the supply of oxygen does not keep pace with the needs of the organisms, it will become a limiting factor. Mechanical aeration, mixing, or the addition of oxygen-containing compounds are effective solutions.

- *Temperature.* Biological growth can occur within a wide range of temperatures, although most microorganisms are active primarily between 10° and 35°C (50° to 95°F). The rate of biochemical reactions in cells increases with temperature up to a maximum, above which the rate of activity declines as enzyme denaturation occurs and organisms either die or become less active.

- *pH Control.* In general, neutral or slightly alkaline pH levels favor biological growth. The optimum pH range for most organisms found in biological treatment systems is between 6.0 and 8.0. Treatment effectiveness is generally not affected by changes within this range; however, pH levels outside of this range can lower treatment performance. The pH of any waste remediated in these processes should be monitored and adjusted during pretreatment, such as during the feed preparation for composting.

- *Microorganisms.* The nature and quantities of organic compounds will affect their biodegradability. In all biological treatment systems, the organisms are naturally subjected to a selection process in which the organisms capable of efficient biodegradation under the given circumstances increase their numbers. (See Subsection 3.2.1 Microbial Ecology and Physiology)

The complex nature of wastes, sludges, and contaminated soils presents a unique challenge in terms of successful remediation and minimization of impacts on other parts of the environment. When determining the applicability of bioremedial technologies, treatment test results from a series of actual site samples, representative of the variety of site conditions, should be evaluated (see Section 3.3, Site Characterization Relative to In Situ Bioremediation, Introduction). In evaluating the potential effectiveness of a technology, the entire treatment train configuration should be considered, including:

- materials handling and preprocessing;
- contaminant destruction, physical transfer, or immobilization; and
- effluent, emission, or residue treatment technologies.

Where possible, the results of previous studies and successful treatment applications should be consulted to aid in the evaluation.

3.5.1 In Situ Bioremediation Technologies

3.5.1.1 Unsaturated Zone

3.5.1.1.1 Land Treatment *Introduction.* In situ land treatment is a managed treatment and disposal technology that entails the application of waste, sludge, or contaminated soil to uncontaminated surface soils at a site and then tilling or plowing the material into the surface soils. This process may also be applied directly to surface soils that have been contaminated by chemical or waste spills. The design and operation of a land treatment facility is based on sound scientific and engineering principles, as well as on extensive, practical field experience.

The objectives of in situ land treatment are:

- biological degradation of organic waste constituents;
- immobilization of inorganic waste constituents; and
- avoiding bioaccumulation of waste constituents that may be detrimental to human health and the environment.

Land treatment uses the assimilative capacity of the soil to decompose and contain the contaminated material in the surface soil layer. The mechanisms of transformation of contaminated soils are illustrated in figure 3.4. Many studies (American Petroleum Institute 1983; Brown, Evans, and Frentrup 1983; Loehr, Martin, and Neuhauser 1992; Loehr et al. 1990) have shown that oil, metals, and other constituents of environmental concern are successfully treated or immobilized by land treatment systems.

Land treatment has been used to treat the following:

- wastes from coal gasification and liquefaction, food processing, leather tanning, paper and pulp production, petroleum refining, and wood-preserving industries;
- sludges and contaminated soils at Superfund sites; and
- municipal wastewater and sludge.

The process has been used under a wide range of hydrogeologic conditions and in the major climatic regions of the United States, Europe, and Canada. Land treatment usually requires minimal management compared with other bioremediation techniques, but does require aeration of the soil,

Figure 3.4
Schematic of an In Situ Land Treatment Process

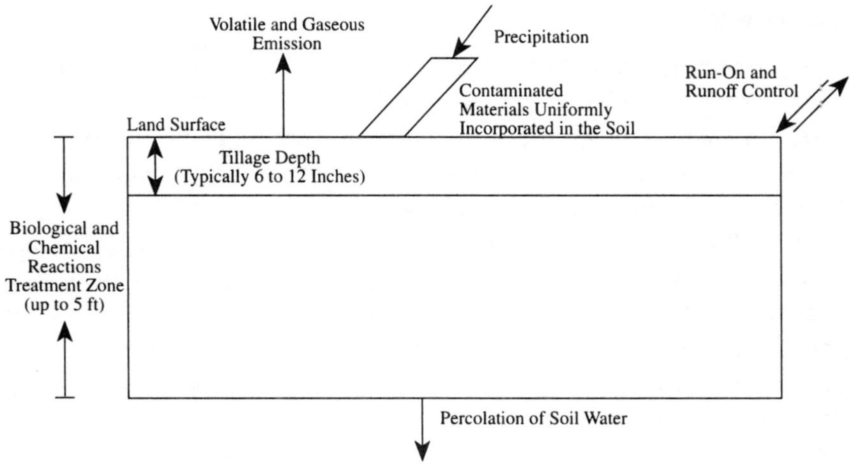

maintenance of optimum soil characteristics (i.e., pH and nutrient balance), and available land.

The soil and waste mixture is managed in a manner that:

- enhances immobilization of waste by soil;
- stimulates degradation of waste by indigenous microflora;
- minimizes volatilization and leaching of waste out of the treatment area; and
- controls surface water runoff.

The mechanisms of immobilization and degradation include sorption, hydrolysis, photolysis, chemical degradation, and biodegradation. Factors that affect biodegradation in land treatment include the type and concentration of the waste, presence of waste-degrading organisms, pH, temperature, and the availability of oxygen, water, and nutrients. Usually, the indigenous soil microflora is stimulated to degrade the wastes; however, microorganisms that have adapted to degrade the contaminants may be added if contaminant-degrading activity is absent in the waste, sludge, or contaminated soil.

Land treatment uses the assimilative capacity of the soil to decompose and contain the applied waste in the surface soil layer (usually the top 15 to 30 cm (6 to 12 in.)). This soil layer is referred to as the zone of incorporation (ZOI). The ZOI and the underlying soils, where additional treatment and immobilization of the applied waste constituents occur, constitute the treatment zone.

Since few organic substances are completely resistant to biodegradation, many organic contaminants can be treated by an in situ land treatment process. But, numerous organic contaminants require a long time to be remediated because they have chemical structures that resist biodegradation. Such compounds include high molecular weight cross-linked polymers, highly branched hydrocarbons, and halogenated compounds. Some of these recalcitrant materials are naturally produced (e.g., coal, tannins, and humic substances), and some are synthetic (e.g., polyethylene). Other compounds are somewhat more susceptible to biodegradation yet are usually classified as nonbiodegradable because of the long time required for appreciable biodegradation to occur. Fortunately, remediation times are long in land treatment units and even seemingly recalcitrant organic conpounds in the applied wastes and sludges frequently are degraded and immobilized.

Organic compounds that are barely water-soluble present difficult biodegradation problems for microorganisms since the organisms and their enzymes must function in an aqueous solution. Therefore, some organic contaminants whose chemical structures should permit biodegradation are slow to decompose because of limited water solubility.

Metals and other inorganic constituents typically are immobilized in the soil and are not degraded or otherwise removed from the site; however, their redox states may be altered. Wastes with high-metal content, such as electroplating sludges, are not prime candidates for land treatment.

Table 3.15 (on page 3.68) arrays the important waste, site, and operational factors that need to be considered in the design and operation of an effective in situ land treatment unit. Other major factors include soil moisture and pH and available nutrients. Water is essential for microbial activity, and nonsporeforming microorganisms will die or remain inactive at very low water concentrations. Alternatively, excessive water levels do not directly harm microorganisms, although they do retard aerobic metabolism by minimizing oxygen transfer in the soil. The aerobic biodegradation of organics in soil is greatest when the moisture content is 50 to 70% of the soil field capacity (Bartha and Dibble 1979). Wet soils also limit site access for waste application and tilling operations.

Although anaerobic degradation occurs in soil, it should be limited for effective land treatment because anaerobic biodegradation is slower and less complete for most organic contaminants. In addition, most metals are more water soluble in a reduced state and more susceptible to leaching.

The optimum pH for soil biodegradation lies between 6 and 8; however, effective biodegradation can occur outside this range. In many land treatment operations, soil pH is kept above 6 to avoid metals migration, as well as to optimize microbial activity. Low soil pH can be modified by adding agricultural lime to the treatment soils, raising the pH to the appropriate range. Microbial growth and metabolism can be reduced by abrupt pH changes; therefore, soil pH should be modified cautiously.

Table 3.15
Land Treatment Design and Operating Factors

Waste Characteristics	
Physical Composition	Salts
Organics	Nutrients
Metals	pH
Site Factors	
Soil Characteristics	*Climate*
Topography	Temperature
Soil texture	Precipitation
Soil moisture content	
Cation Exchange Capacity	*Hydrogeology*
Soil pH	Depth to seasonally high water table
Soil microorganisms	Depth to usable aquifer
Nutrients	Proximity to surface water
Operational Factors	
Waste Application	*Soil Amendments*
Organic loading	Nutrients
Hydraulic loading	Moisture
Frequency of application	pH control
Method of application	
	Monitoring
Storm Water Management	Waste Characteristics
Run-on/Runoff control	Soil
	Leachate
Waste Incorporation	Vegetation (if grown)
Depth of incorporation	Runoff (if any)
Frequency of cultivation	

Available oxygen is critical for any form of aerobic bioremediation. Soil microorganisms use oxygen transferred to soil water from the atmosphere. Oxygen availability is a function of:

- the amount of void space in the soil;
- the partial pressure of oxygen in the soil atmosphere;
- the oxygen transfer rate from soil atmosphere to soil water; and
- the rate at which soil microorganisms are using the available oxygen.

It is possible to maintain aerobic conditions in a land treatment unit by:

- keeping the soil unsaturated;
- moderate tilling;
- avoiding unnecessary compaction (heavy trucks, etc.); and
- limiting the loading of rapidly-biodegradable matter so that oxygen demand does not exceed the oxygen transfer rate.

Soil texture and permeability also affect aerobic conditions. Sand or loam is conducive to aerobic bioremediation, while heavy clay is not. Sandy soils permit rapid infiltration of liquids and oxygen, but also may allow pollutant migration. Clayey soils provide better containment, but can reduce infiltration of irrigation water and precipitation. Clayey soils also reduce air exchange, potentially causing anaerobic conditions.

Site Preparation and Equipment. Site preparation for in situ land treatment includes: (1) removing site vegetation, (2) separating the sites into several plots, and (3) constructing an elevated roadway/dike around the facility to control surface run-on and runoff and permit site access. Many land treatment sites are graded to promote surface drainage and collect runoff. Slopes in excess of 5% are generally not recommended because of erosion and runoff control problems. A 1 to 2% grade is common, providing controlled runoff and preventing ponding of added water.

Site preparation activities vary depending on the existing site characteristics and regulatory requirements. Following are some examples:

- removing trees and rocks for site access and ease of tilling;
- digging drainage ditches to intercept seasonally-high perched water table and runoff;
- adjusting soil pH for low pH soil; and
- contouring, terracing, and grading to intercept and divert off-site run-on, contain on-site runoff, and provide access to the site.

During operation, the land treatment unit may need fertilization and pH control (by liming of the soil). The amount and frequency of lime application depends on the pH and the buffering capacity of the soils, but typical ranges are from 1.8 to 3.6 tonne (2 to 4 ton) of lime per acre every 1 to 3 years on acid soils to maintain near neutral soil pH (American Petroleum

Institute 1983). Generally, soil monitoring data are used to determine the lime requirement.

Nutrient applications (nitrogen, in particular) may enhance organic degradation rates in the initial period, but may not be necessary in subsequent years. Over-application of nitrogen fertilizers may result in excessive leaching of soluble nitrates. Other nutrients, in addition to those applied, are available at land treatment sites. Many agricultural soils have a high reserve of phosphorus. Mineralization of organic nitrogen in the soil, the applied waste, and sludges may provide a significant nitrogen input and can recycle applied nutrients after repeated waste applications.

The waste, sludge, or contaminated soil is applied uniformly and tilled in at a usual depth of 15 to 30 cm (6 to 12 in.), although depths of up to 46 cm (18 in) are practical with heavy tilling equipment. The cost of the equipment and the available land area will affect the selection of the tilling depth. After the application, distribution, and incorporation of the waste, additional tilling generally is performed to:

- increase the available area for soil microbial contact with contaminants;
- maintain aerobic and homogenous conditions; and
- maintain a loose, moist, and well-mixed soil to maximize organic contaminant decomposition.

Performance. For hydrocarbon-contaminated wastes and soils, removal of total petroleum hydrocarbons (TPHs), as measured by gas chromatography (GC) methods, will be in the range of 90 to 99%. In some instances, treatability studies have indicated removals of the following nature: PAH such as naphthalene, acenaphthene, and fluorene – 80-95%; higher ring PAH – 30-70%. Field studies have achieved removals in the same range when proper conditions for biodegradation have existed. Loss rates for organic constituents are typically described by the "half-life" of the organic compound. Compounds with higher aqueous solubilities have relatively shorter half-lives than less soluble compounds. The average half-life for typical oil and grease found in refinery wastes and oily sludges in land treatment units ranges from 50 to 150 or more days. Removal or loss rates vary widely and depend on the type of hydrocarbons and site-specific factors affecting the rates of biological activity.

Most of the effort in defining degradation rates for hydrocarbon-contaminated materials has been spent evaluating loss rates for volatile and semivolatile compounds: BTEX — benzene, toluene, ethylbenzene, xylenes—and PAHs. The PAHs are of interest because several are suspected carcinogens. The BTEX compounds are removed (lost) rapidly by a combination of volatilization and biodegradation; the half-lives are generally less than a week.

The primary mechanisms influencing the fate of PAHs are volatilization, adsorption/desorption, and biological oxidation. In solids (e.g., soil and sludge), no significant volatilization or biological oxidation of PAHs will occur unless they desorb from the solid phase. Therefore, PAH adsorption/desorption will be the rate-limiting step in the fate of PAHs. Only the two- and three-ring PAHs have a high volatilization potential. The more complex PAHs can be biodegraded, but have long half-lives of weeks to months.

The owner or operator of a land treatment unit must establish a program to ensure that organic and inorganic contaminants applied are degraded, transformed, or immobilized within the treatment zone. Federal and state regulations define the principal components of an in situ land treatment facility as:

- the wastes to be applied;
- the design and operating measures necessary to maximize degradation, transformation, and immobilization of the waste constituents; and
- an unsaturated zone monitoring program.

Before applying waste, the owner or operator must demonstrate that constituents can be completely degraded, transformed, or immobilized in the treatment zone. Generally, a treatment demonstration is required to establish that the operating practices at the site will protect human health and the environment, considering the characteristics of the waste, soil, and climate at the site. The treatment demonstration will be used to determine permit requirements and operating principles. The U.S. EPA has published a permit guidance manual governing the treatment demonstration (US EPA 1986).

In situ land treatment units also must have a groundwater monitoring program to detect and correct any groundwater contamination. The permit guidance manual (US EPA 1986) prescribe the requirements for runoff and run-on controls, growth of food chain crops on such sites, closure and post-closure care, and record keeping.

Advantages/Disadvantages. One advantage of in situ land treatment is that the organic and inorganic contaminants are reduced in concentration and/or immobilized and not simply disposed. Other advantages include low capital and operational costs, low-to-moderate manpower and maintenance requirements, and no long-term liability ensuing from off-site disposal. If the unit is overloaded or poorly operated, however, organic and inorganic compounds might leach below the land treatment unit. Other disadvantages are the need for a large land area and limited control over volatile organic loss to the atmosphere.

3.5.1.1.2 Bioventing *Introduction.* Bioventing is the use of induced air movement through unsaturated soils, with or without nutrient addition, to reduce soil contamination through biodegradation (Hinchee 1993; Van Eyk and Vreeken 1989b). The process stimulates the indigenous microorganisms to convert organic contaminants, such as petroleum hydrocarbons, to less hazardous substances, especially carbon dioxide and water.

Most bioventing systems have used air recovery wells (see figure 3.5 on page 3.73), such as those used in in situ vapor recovery systems, to move air and thus provide oxygen for the indigenous bacteria. Many systems that were designed as in situ vapor recovery systems to physically remove the contaminants were found to stimulate biodegradation as well; these were subsequently relabeled as bioventing systems (Bennedsen, Scott, and Hartley 1987).

Some systems have been designed with air injection wells either alone (figure 3.6 on page 3.74) or in conjunction with air recovery wells (figure 3.7 on page 3.74). The use of air injection can be as effective as air recovery in terms of providing air for biodegradation. Air injection should not be used near buildings because it could result in exposure or explosion hazards. It can be used to treat heavy petroleum blends in areas away from buildings and utility trenches, and in conjunction with an air recovery system.

Systems that are designed primarily to promote biodegradation, as opposed to physical removal, will have the air recovery wells located outside the most heavily-contaminated area (figure 3.8 on page 3.75) and incorpo-

Figure 3.5
Bioventing Design to Maximize Recovery of Volatile Compounds

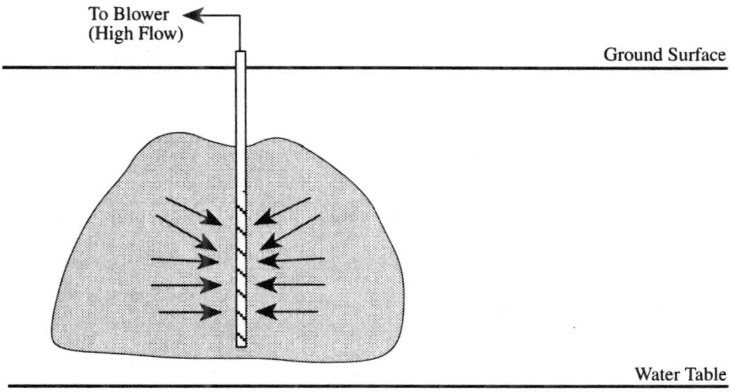

rate significantly lower air flows than systems designed for physical removal. This approach has the advantage of requiring less offgas treatment because it focuses on destroying the contaminants rather than transferring them to a different medium.

Nutrient addition is not always incorporated in bioventing systems. In some cases it is not necessary, and in others it may be very difficult to accomplish. Where the soil types and site infrastructure permit, aqueous nutrient solutions can be percolated from the surface. In low-transmissive soils or where surface access is restricted, percolation of nutrients may be difficult or infeasible. Attempts to introduce gaseous phase nitrogen sources have not yet been successfully demonstrated.

Scientific Basis. The scientific basis for this technology has the following aspects:

- the metabolic capabilities and requirements of microorganisms;
- the dynamics of air flow through unsaturated soils;
- the adsorption and desorption of the organic contaminants on soils;

Figure 3.6
Bioventing Design to Optimize Biodegradation Using Air Injection Only

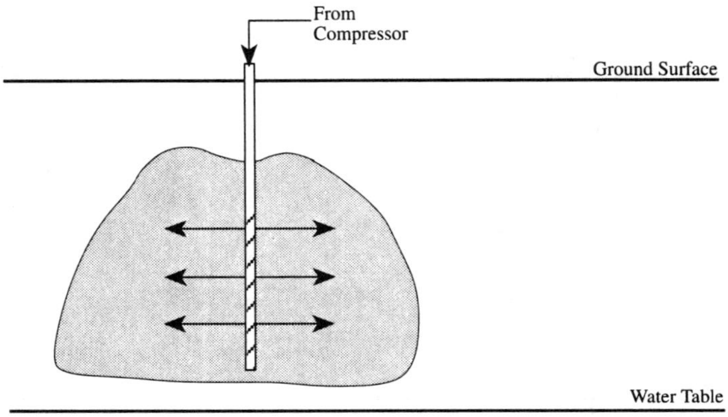

Figure 3.7
Bioventing Design to Optimize Biodegradation
Using Air Recovery and Injection

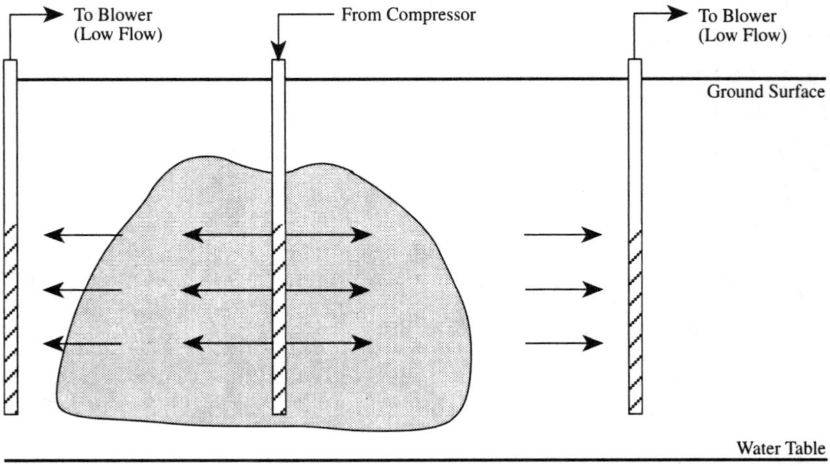

- the transport of nutrients through soils; and
- the volatility of the organic contaminants.

The microbiology of bioventing is basically the same as that applied during and understood from years of biological treatment of wastewaters and land treatment applications. The practical basis for implementation of the technology has resulted from development of in situ vapor recovery and bioremediation soil-pile technologies.

As explained in Subsection 3.2.1, Microbial Ecology and Physiology, a variety of common aerobic soil bacteria can use a variety of classes of organic compounds as food and energy sources. Other microorganisms may produce enzymes capable of at least partially degrading some compounds, while utilizing different classes of compounds as their primary food and energy sources.

The stoichiometric oxygen requirements for the conversion of hydrocarbons to carbon dioxide is approximately 3 kg (6.6 lb) of oxygen for 1 kg (2.2 lb) of biodegradable organic matter. In wastewater treatment systems, actual oxygen demand is more typically 0.8 to 1.7 kg (1.8 to 3.7 lb) of oxy-

Figure 3.8
Bioventing Design Optimize Biodegradation Using Air Recovery Only

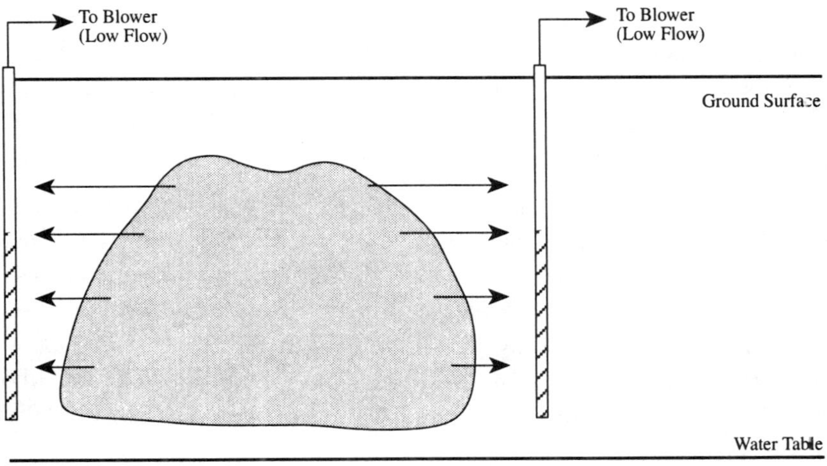

gen per kg (lb) of degradable carbon (Eckenfelder 1967). The actual oxygen requirement depends on the degree of mineralization of the contaminant and the efficiency of the oxygen supply system. If local anaerobic conditions exist, anaerobic degradation of the intermediate products may serve to lower the effective oxygen demand.

Nutrient requirements other than nitrogen and phosphorus are relatively small and in nearly all cases can be met by the existing minerals in the soils. Nitrogen and phosphorus requirements for cell production have been estimated at 4.3 kg (9.5 lb) of nitrogen and 0.6 kg (0.13 lb) of phosphorus per 100 kg (220 lb) of biological oxygen demand (BOD). Actual demand in any system will depend on the extent of direct oxidation of the organic matter versus cell growth, rate of recycle of nutrients from dead cells, and existing sources of nutrients including bacteriological nitrogen fixation.

The dynamics of air flow through soils are well understood, but have not been studied as thoroughly as those of groundwater flow (Wilson, Clarke, and Clarke 1988). The permeability of soils to air is two to three orders of magnitude greater than the soils' permeability to water (Wilson and Ward 1986). As a result, pressure differences will cause air to flow through the interstitial spaces between the soil pores, provided the spaces are not filled with water or other liquids. The ease of flow is dependent on the soil permeability and moisture content, and thus air will flow more readily through sands than through clays. Diffusion of oxygen into pore spaces is important and contributes substantially to the distribution of oxygen. The air flow through soils has been thought to follow Darcy's law and has been modeled and predicted on this basis. Recent work by Wilson et al. (1992) suggests that Darcy's law does not hold in many cases (where the Reynold's number exceeds four or, perhaps, one). Several models for in situ vapor recovery have been developed and used for the physical removal of volatiles from unsaturated soils (Van Eyk and Vreeken 1989a).

Because air is easier to move through soils than water and because air has a higher oxygen-carrying capacity than water, it is possible to deliver a relatively large amount of oxygen through soils even of moderate or low permeability. For instance, a very modest air flow rate of 0.3 m^3/min (10 ft^3/min) will introduce 100 kg (220 lb) of oxygen per day or 38,000 kg (83,790 lbs) per year, equivalent to enough oxygen to convert hydrocarbons to carbon dioxide at a rate of 15.9 kg (35 lb) per day or 5,670 kg (12,500 lb) per year.

Organic compounds with a vapor pressure greater than approximately 0.1 mm at normal atmospheric conditions are sufficiently volatile to be recovered through in situ vapor recovery under favorable conditions. The rate at which a particular compound can be volatilized and moved through a particular soil depends on whether the compound is present as a free phase, adsorbed to the soil particles or humates associated with the soils, or dissolved. The tendency of organic compounds, which are neither adsorbed within the soil particles or dissolved, to volatilize can be predicted from their vapor pressures. Volatilization of compounds that are in the dissolved phase can be predicted from the Henry's Law constant. Use of either parameter alone can be misleading; in general, the Henry's Law constant is highly preferred. An Apparent Henry's Law constant, or lumped partitioning coefficient, which must be determined experimentally, may be preferred for use in computer models of contaminant attenuation (Wilson, Clarke, and Clarke 1988).

Once volatilized, compounds can readsorb as they move through the soil matrices. Adsorption and readsorption on soil particles brings the volatile organic compounds in contact with the soil bacteria, permitting biodegradation. Ostendorf and Kampbell (1990) have documented a hydrocarbon-vapor transport model that takes into account soil and organic constituent properties, biodegradation rates based on Michaelis-Menton kinetics, and air flow through unsaturated soils. The model was tested in the laboratory and at two test sites and resulting data were compared to data generated by Marley and Hoag (1984) and by Thorton and Wootan (1982). The model's data suggested that volatile constituents of aviation gasoline can be reduced effectively in soil containing adapted microorganisms and sufficient nutrients.

Nutrients may become available to the bacteria through minerals present in the soils, addition of a nutrient-enriched leachate, or in the case of nitrogen, through biological nitrogen fixation (DuPont 1992a). Attempts to add nitrogen as ammonia gas have not been successful (R.E. Hinchee, private communication). Ammonia gas is adsorbed to soils within a few inches of the injection point. This rapid adsorption has been taken advantage of by farmers who introduce the gas a few inches below the surface of the soils. Typically, loss of ammonia vapors to the atmosphere cannot be detected readily.

The feasibility of nutrient addition to unsaturated soils is dependent on both access to the surface and on the affinity of the soils for the individual nutrient components. If the surface is covered, nutrient addition will be more cumbersome. Nutrients are usually added as an aqueous solution that is allowed to percolate through the soils. Addition of a nutrient solution may be followed by addition of unamended water or by fortuitous rainfall. The most available nutrient sources, such as ammonium and phosphate, can be adsorbed by soil particles; thus, their movement may be retarded through soils in an aqueous solution. Nitrate has little affinity for soils and, therefore, its movement in the unsaturated zone is retarded only by biological consumption and dilution with residual moisture. The degree of adsorption of ammonium and phosphate sources is greater in clayey or silty soils than in sandy soils. Among the common phosphate sources, orthophosphate is typically adsorbed less than are the polymeric phosphates such as tripolyphosphate (Norris and Subramanyam 1992) and trimetaphosphate (Aggarwal, Means, and Hinchee 1992). While adsorption retards the movement of nutrients through the soils, it also causes the nutrients to be retained on the soil surfaces where they will be available for the bacteria.

Nitrogen fixation under anaerobic conditions has been observed. Whether or not nitrogen fixation can provide significant quantities of nitrogen for contaminant biodegradation at a spill site is unknown. A recent study by DuPont (1992b) concluded that nitrogen fixation requires the presence of a readily-metabolizable substrate, such as glucose, before the process can provide a significant source of nitrogen in bioventing applications. Others have speculated that nitrogen fixation may be significant at old spill sites, and two recent field studies indicated that nutrient addition is not always necessary or beneficial. These studies do show, however, that moisture management is important (DuPont 1992b).

In summary, what is known about metabolic processes and requirements, movement of air through soils, and transport of inorganic nutrients, supports the concept of bioventing as a remediation process for unsaturated soils containing petroleum hydrocarbons and other biodegradable constituents. The actual need for nutrients under field conditions is not yet well understood, and some available field and laboratory data appear contradictory. The issue is further complicated by the difficulty of distributing nutrients under some conditions.

System Designs. All system designs incorporate some method of introducing oxygen to the unsaturated soils and may include a method of nutrient addition. Most systems have used wells to recover or inject air; however, for shallow water table sites, trenches may be preferable, provided that the excavated soils will not be listed as hazardous waste. Figure 3.6 (on page 3.74) shows a typical design for a vapor extraction system using wells. The wells are placed in the contaminated zone and screened over some interval depending on the distribution of the contaminant, the soil type distribution, and whether the surface is covered by an impermeable layer. In uncovered sites, the greatest radius of influence of a well, particularly within the capillary zone, is effected with wells that are screened over a narrow interval located just above the water table (or impermeable zone) or near the bottom of the contaminated interval. For covered sites, the well can be screened over a longer interval and extend closer to the surface.

When the contamination extends into the saturated zone, the ability to supply oxygen much more rapidly and inexpensively through air movement (as opposed to water circulation) should be considered. If the water table is at least 3.1 m (10 ft) below the surface, the well screen may extend below the water table to take advantage of the exposure of additional soils during periods when the water table is lowered. The water table can be lowered by pumping out groundwater in the vicinity of the treatment area. In highly-permeable soils, very large volumes of water will need to be recovered and treated. In formations with a high-clay content, the volume of water required to dewater the zone of interest will be much smaller than in more permeable aquifers; however, the soils might not drain rapidly enough to permit sufficient permeability for air movement into the exposed soils. The applicability of dewatering needs to be evaluated by hydrogelogists experienced in dewatering sites.

If the primary purpose of a system is to promote biodegradation rather than physical removal of volatile constituents, the air extraction wells should be located on the periphery of the contaminated area as shown in figure 3.8 (on page 3.75). This will increase the flow path of the air and incorporate a larger volume of soil into the treatment zone. Maximizing the volume of soil in the treatment zone makes more bacteria and native nutrients available for contaminant biodegradation.

Locating the wells on the periphery of the treatment zone and not centrally located within the zone will increase and possibly double the number of wells required to sufficiently aerate the treatment zone. However, placing wells in the contaminated zone will result in offgassing of contaminants that will require secondary treatment. Installation of more wells is less expensive than secondary treatment of volatilized contaminants.

Using low air extraction rates, volatile contaminants generally will have more time to be biodegraded before being physically removed from the subsurface. Furthermore, constituents volatilized in the more highly-contaminated areas may be readsorbed and biodegraded as they pass through the less contaminated areas as shown in figure 3.8 (on page 3.75).

As with vapor extraction systems, it may be advisable to operate the system intermittently. Because oxygen demand decreases as biodegradation proceeds, the rate of oxygen supply can be reduced continually as long as at least 2% oxygen is maintained in the contaminated zone (DuPont 1992a). Once the contaminants have been degraded to an appropriate level, the rate of oxygen supply can be reduced by cycling the recovery wells or by operating the recovery wells on a rotating schedule. The latter approach can result in the use of a smaller blower and reduced demand for offgas treatment. Rotating wells and adjusting airflows also reduces the chance that dead spots, regions of little or no flow, will occur.

Air injection wells can be used in conjunction with air recovery wells. Air can be injected into either the saturated or unsaturated zone. Because air injection may cause volatile constituents to migrate away from the injection point, these systems require more care in their design than those in which air is simply extracted.

Injecting air into the unsaturated zone may provide the greatest benefit by screening the injection wells in areas of low permeability, while screening the air extraction wells in the more permeable zones.

Soil moisture levels are an important factor. Biodegradation rates are enhanced by high-moisture content. But, water-saturated soils inhibit the distribution of air and, therefore, oxygen. One rule of thumb is to maintain moisture levels between 40 and 60 percent of field saturation. Designing and operating bioventing systems to maintain low-air flows minimizes loss of moisture. In fine soils, it may be advantageous to remove some of the moisture initially to allow adequate movement and distribution of oxygen.

Despite the well-documented requirements for nutrients in biological wastewater treatment and in laboratory treatability studies, etc., the need for nutrients in bioventing still is not well understood. The maximum amount of nutrients that should be required is a ratio of 100:10:1 or 2 of C:N:P. Because nutrients are recycled and not all of the carbon is used for cell growth, the requirements are less than the above by a least one-half and probably by greater than one-fourth. Furthermore, nutrient addition is not always readily effected. Currently, it is effected through the addition of aqueous solutions of nitrogen- and/or phosphorus-containing compounds, which also add moisture. Thus, nutrient and moisture maintenance must be coordinated.

At sites where the surface above the contaminated zone is not covered, dilute nutrient solutions or granular nutrient blends can be added to the surface and allowed to percolate through the soils. Additional water can be supplied through a combination of spraying and fortuitous rainfall to transport the nutrients deeper into the formation.

At sites where the surface is covered with asphalt or cement, nutrients may be added just below the surface provided there is a sufficient gravel or sand layer below the asphalt or cement. When a gravel or sand base is not present or where contamination is present at depth, nutrients and moisture may be added through a series of wells or trenches. If nutrients are added beneath or close to a building, care must be taken not to adversely affect the load-bearing capabilities of the soils.

The selection of nutrient sources is also important. Since little is known about nutrient requirements during bioventing, it is difficult to predict with certainty which nitrogen and phosphorus sources will be most effective in stimulating microbial activity. It is clear, however, that nitrate percolation into the formation will be less retarded by adsorption on soils than ammonium percolation. Furthermore, orthophosphate will be retarded less than tripolyphosphate, and, in most soils, orthophosphate will be retarded more than ammonium. Clearly, it will require less water to transport nitrate to a given depth than will be required to transport ammonium. On the other hand, some sorption of the nutrients to the soil may be beneficial. In some cases, nitrate might be flushed through the soils and unavailable to the bacteria. In addition, nitrate may be transported into groundwater and reach levels that exceed drinking water standards (10 mg/L as nitrogen).

Flushing nutrients through the soils also creates some potential for transporting contaminants from the unsaturated to the saturated zone. If the saturated zone is already contaminated with the same constituents, whatever remediation system is implemented in the saturated zone should accommodate the organic contaminants transported into that zone. Operation of the air recovery system before nutrient addition should reduce the potential for vertical transport of contaminant constituents by oxidizing and/or physically removing the most mobile constituents.

Impact. Bioventing has its most direct impact on the unsaturated zone. The process will remove and/or degrade a number of organic contaminants. Nutrients that are added, but not utilized by the microorganisms, will cause some increase in nutrient concentration. Some changes in inorganic species may occur, such as an increase in the oxidation state of metals such as iron. Depending on how the system is operated, soil moisture may increase or decrease.

Bioventing may prevent or minimize contamination of groundwater in removing the soluble constituents from the unsaturated zone and preventing their migration into the saturated zone. Bioventing also may remove and/or degrade constituents within the capillary zone and exposed soils when the water table is low.

Air quality can be negatively impacted if large quantities of volatile constituents are discharged to the atmosphere. Operating at low air flow rates and/or use of offgas treatment protects air quality. In some cases, especially where there are significant concentrations of volatile compounds near the surface, the bioventing process will reduce the mass of volatiles reaching the atmosphere.

Pre- and Posttreatment Requirements. There are no special pre- or post-treatment requirements other than site delineation, post-treatment sampling, and getting permits.

Design Data, Unit Sizing. To design an effective bioventing system, it is necessary to determine the distribution of the contamination, whether the contaminants are biodegradable and/or sufficiently volatile, the soil properties and depth to water, and the site infrastructure.

Unfortunately, there is never as much site characterization data as desired. As a result of the inherent complexity of subsurface conditions and the cost of thoroughly defining them, it is usually necessary to design some flexibility into the system. However, it is necessary to have sufficient data to achieve the goals of the design effort.

The identification of all constituents (or mixtures) of interest is necessary in order to know whether the process has a reasonable chance of succeeding. The biodegradability and/or ventability of all constituents of interest has to be established from previous experience, published data, or laboratory treatability studies.

The mass of contaminants present in the area to be addressed must be estimated in order to anticipate the potential nutrient and offgas treatment requirements, as well as to estimate remediation times.

The soil types and distribution and the depth to water must be determined in order to conceptualize how a system might be designed and to determine whether there is sufficient potential to warrant pursuing this option. Detailed designs typically require field tests be conducted. The complexity of the field test will depend on the site conditions, including the volume of soil to be treated. The larger, more complex sites justify more extensive pilot studies than do small sites. In small sites, such as service stations, a large portion of the contaminated zone is located close to the tank pit and/or underground utility trenches. Consequently, a test at any one location would not be representative of any other location. It may be best in such cases to design the air extraction wells in a conservative fashion and be prepared to add a few additional air extraction wells as necessary.

Soil gas surveys can be conducted initially to provide a general sense of the distribution of volatiles and biodegradable substances. Traditional soil gas testing can be augmented to provide an approximation of soil permeability and thus the ease of providing oxygen through aeration. Early use of soil gas techniques may improve the efficiency and lower the cost of soil boring to delineate the contamination.

Laboratory tests can be conducted to determine the feasibility of bioventing, estimate the rate and extent of biodegradation, and provide data for use in designing engineering tests. Plate counts, pH, and results of respirometer tests, often are used to identify conditions inhibiting biodegradation. Microcosm studies have been conducted to determine the rate and extent of biodegradation, the effect of added nutrients on soil permeability, and the nutrient requirements of the microorganisms. Unfortunately, these tests frequently do not enable accurate prediction of field results. Tests to determine the ease of nutrient percolation through the soils can provide qualitative information on the ease of nutrient transport and can determine the nutrient sources that will move most readily through the soils.

In situ respirometry tests can determine the existence of ongoing biodegradation (Hinchee and Ong 1992). Such tests may include the use of tracer gases or the equivalent of aquifer pump tests to determine the potential for induced air flow through pilot-test areas. The former can be as simple as analyzing soil gas samples in both contaminated and uncontaminated zones for oxygen and carbon dioxide levels. Low-oxygen levels and high-carbon dioxide levels suggest that biodegradation is occurring. Potential rate data can be obtained by monitoring soil gases in conjunction with intermittent air recovery. The data in table 3.16 indicate that hydrocarbon biodegradation rates can vary from 0.2 to 20 mg/day per kg of soil (2.0×10^{-7} to 2.0 x

Table 3.16

Comparison of Biodegradation Rates Obtained by the in Situ Respiration Test with Other Studies

Site	Scale of Application	Respiration Contaminants	Estimated Rates (% O_2/hour)	Biodegradation Rates	References
Various (8 locations)	In situ respiration tests	Various	0.02 - 0.99	0.4-19 mg/kg/day	Hinchee and Ong 1992
Hill AFB, Utah	Full-scale, 2 years	JP-4 Jet Fuel	up to 0.52	up to 10 mg/kg/day	Hinchee et al. 1991
Tyndall AFB, Florida	Field pilot, 1 year and in situ respiration tests	JP-4 Jet Fuel	0.1 - 1.0	2 - 20 mg/kg/day	Miller 1990
Netherlands	Undefined	Undefined	0.1 - 0.26	2 - 5 mg/kg/day	Urlings et al. 1990
Netherlands	Field pilot, 1 year	Diesel	0.42	8 mg/kg/day	van Eyk and Vreeken 1989b
Undefined	Full-scale	Gasoline and Diesel	--	50 kg/well/day	Ely and Heffner 1988
Undefined	Full-scale	Diesel	--	100 kg/well/day	Ely and Heffner 1988
Undefined	Full-scale	Fuel Oil	--	60 kg/well/day	Ely and Heffner 1988
New Zealand	Pilot-scale/ Full-scale	Diesel Spent Oil	--	0.2 - 20 mg/kg/day	Hogg et al. 1992

a. Rates reported by Hinchee et al. (1991) were first-order with respect to oxygen; for comparison purposes, these have been converted to zero-order with respect to hydrocarbons at an assumed oxygen concentration of 10%.
b. Rates reported as oxygen consumption rates; these have been converted to hydrocarbon degradation rates assuming a 3:1 oxygen-to-hydrocarbon ratio.
c. Units are in kilograms of hydrocarbon degraded per 30 standard ft^3 per min (scfm) extraction vent well per day.

10^{-5} lbs/day per lb of soil). This type of rate data can be used to estimate potential remediation times based on the initial contaminant levels. Table 3.17 lists calculations made in this manner along with calculations based on stoichiometric oxygen requirements and air-flow rates.

Table 3.17
Potential Times for Contaminant Reduction[1,2]

Initial Concentration (µg/kg)	10 mg/kg/day Reaction Rate	1 Pore Volume Per Day Utilization[3]	3 Pore Volumes Per Day Utilization[3]
	Days		
500	40	20	8.7
1,000	90	45	17
5,000	490	245	82
10,000	990	495	165
20,000	1,990	995	332
50,000	4,990	2495	832

1. Assumes regulatory approved final concentration of 100 ppm; no abiotic loses, and a 3:1 oxygen-to-carbon utilization.
2. From Hinchee and Ong 1992.
3. Based on utilization of all oxygen in indicated volume of air.

Oxygen concentration is a more reliable indicator than carbon dioxide concentration because of the complex behavior of carbon dioxide with respect to adsorption by calcium minerals and solution/dissolution from groundwater and soil moisture. Carbon-13 measurements can distinguish between carbon dioxide resulting from biodegradation of petroleum hydrocarbons and that released by dissolution of minerals.

Pilot and feasibility tests are generally conducted using one or more air extraction wells and several air monitoring points located at various directions and distances from the extraction well and at varying depths. Pressure changes in the monitoring points are determined as a function of air flow and vacuum applied to the extraction well(s). These data can be used to calculate soil permeabilities and plotted manually to estimate the radii of influence of extraction wells. Preferably, a computerized modeling program is used to determine the air-flow paths and radii of influence of the tested extraction wells based on pilot-test data. A treatability protocol developed by the U.S. Air Force and tested at over 100 sites is available (Hinchee et al. 1992).

Information for Consideration. Information required to evaluate and employ bioventing includes the properties of all contaminants of interest, the site hydrogeology, the presence of any underground utilities, foundation designs, regulatory requirements, including permissible soil concentrations and air discharge limits for the constituents of interest, and permit requirements.

The primary contaminants of interest must be biodegradable under aerobic conditions. Volatile, nonbiodegradable constituents can be treated, but will increase the cost of offgas treatment and may require larger blowers or different well configurations and, possibly, longer treatment times. It is important that all constituents be at levels below that which would be toxic to the microorganisms. If compounds that are neither biodegradable nor ventable are also present above the regulatory levels or other goals, bioventing alone will not be effective.

The type of soil and its permeability affect the ease of air flow through the soils. Bioventing is much more difficult in soils with low permeability, such as clayey soils, particularly when moisture levels are sufficient to fill nearly all of the pore space. (To some extent, operation of an air recovery system will remove moisture and provide better air distribution.) Shallow soils (less than 10 ft deep) can experience significant temperature variation during the year. Thus, the rate of biodegradation in these soils due to bioventing can also vary with time during the year. The average annual rate can be maximized by employing soil-warming techniques such as percolation of warm water through the contaminated zone, or by burying heating tape (Leeson et al. 1994).

The depth to the water table or other gas impermeable layer will affect the radius of influence of an air extraction well and the decision to use vertical wells or trenches. Unless the surface is covered with cement or asphalt, a depth to water of at least 3.1 m (10 ft) is necessary to avoid using a large number of wells or vapor recovery trenches. When the surface is covered, a single recovery well may have a very large radius of influence, provided that the cover has maintained its integrity.

The more stringent the cleanup goal, the larger the area that must be treated and the longer the system must be operated. This is particularly true for sites contaminated with high levels of heavier petroleum hydrocarbons. Typically, lighter hydrocarbons can more readily be converted than heavier hydrocarbons.

Technology Variations. Many variations of bioventing can be devised to move air through the contaminated soil. These include air recovery wells and trenches combined with air injection wells, trenches, and sparge points. Injected air can be recycled from the air recovery blower with or without offgas treatment. Examples of some generic approaches are shown in table 3.16 (on page 3.84) and in figures 3.5 to 3.8 (on pages 3.73 to 3.75).

One should take advantage of site conditions rather than try to forcefit a familiar system on a site. For instance, air recovery wells may not be appropriate for sites with an open surface and a shallow water table. Alternatives should be considered, including air sparging to provide oxygen or, if the contamination is shallow enough, land treatment or excavation with some form of treatment such as a soil pile. Selection of excavation should take into account the possibility of uncontrolled releases of VOCs.

For soils contaminated with nonvolatile biodegradable constituents, air injection through wells, trenches, or sparging may be remedial options and could represent the lowest cost alternative. Air injection systems may also be considered where there is sufficient soil above the contamination to act as a biofilter.

Cost. Where applicable, bioventing has the potential to be a low-cost remediation method. The cost of bioventing will depend not only on the site location and conditions and cleanup goals, but also on the particular design employed. Despite its similarities to in situ vapor stripping, the cost of bioventing should be significantly less because the need for offgas treatment is reduced and the total volume of air moved through the soils is lower. For bioventing of heavier petroleum hydrocarbons, particularly if substantially weathered, offgas treatment may not be required at all.

It is assumed here that a bioventing system design will maximize biodegradation relative to physical removal. Thus, systems should provide air at rates only slightly exceeding those required to allow the maximum rate of respiration. The designs should also place wells outside the areas of high contamination to increase the amount of soil available to support biodegradation. These measures, while promoting in situ bioremediation, limit case history information because most bioventing systems have been designed as in situ vapor recovery systems with biodegradation as an unquantified side benefit.

It has been estimated that venting of gasoline requires approximately 100 L of air per g (1,660 ft^3/lb)of gasoline. Assuming a 20% efficiency, the system would have to operate long enough to provide 500 L of air per g (8,000 ft^3/lb) of gasoline. Stoichiometric calculations indicate that approximately 10 L (0.3532 ft^3) of air is required to provide sufficient oxygen to convert 1 g (0.002 lb) of gasoline to carbon dioxide and water. Assuming a 20% efficiency of the introduced oxygen, 50 L of air per g (800 ft^3/lb) of gasoline would be required under a bioventing mode. Many assumptions underlie these calculations, but they serve to illustrate that the demands on the air recovery system will be much less for bioventing than for vapor recovery systems.

In a companion monograph[1], the costs for in situ vapor recovery technology are estimated for a hypothetical site. For vapor recovery, total costs are highest for systems designed for the greatest rate of remediation because of the high-capital costs. Costs decrease rapidly as the design de-emphasizes speed and becomes less capital intensive, and then gradually increase as long-term operation and maintenance costs begin to offset lower capital costs. Bioventing systems resemble the less capital intensive vapor recovery systems, except that offgas treatment capital and operating costs generally will be less.

When considering costs, it is also assumed that site delineation has been completed and that groundwater dewatering is not required. The costs for bioventing can be divided into engineering, capital, and operating/monitoring. Engineering costs include field tests to determine the radius of influence and air-flow characteristics of the soils. This allows the determination of well spacing and screening intervals. Field tests using existing wells and soil gas probes can cost under $10,000 for small homogeneous sites. Costs can reach $100,000 for larger, more complex sites where field respirometry and biodegradation assays are conducted.

1. See Innovative Site Remediation Technology: Vacuum Vapor Extraction—Ed.

System installation will depend on the depth of soils being treated and drilling conditions. Well installation costs will range from \$2,000 to \$5,000 per well for most sites. Blowers, controllers, electrical, equipment building, security, etc. should range from \$20,000 to \$50,000. Costs will be higher for tighter soils because the well spacing will have to be closer than for sands and gravel.

A significant cost element is the offgas treatment system. Hinchee (1993) estimated that bioventing costs might increase from \$13/m^3 (\$10/yd^3)to as high as \$52 to \$78/m^3 (\$40 to \$60/yd^3) if offgas treatment is included. Offgas treatment can range from a few thousand dollars for two carbon canisters to \$100,000 for a catalytic incinerator. If offgas treatment is included, monitoring costs will also increase.

Operating costs can be minimal if the monitoring program is managed well. High quality equipment and proper design will minimize maintenance and travel to the site for routine monitoring. Monitoring can be done primarily with portable equipment to measure flow rates, pressures, and concentrations of oxygen, carbon dioxide, and organic vapors. This approach will limit laboratory costs to those for periodic confirmatory sampling for organic vapors. Soil sampling is another cost. Overall operation and maintenance costs, depending on the size of the operation, may range from \$10,000 to \$50,000 per year.

Case histories with the most reliable and thorough technical information do not usually provide the best cost information. Cost information from specific sites is not always in a form that allows direct cost comparison of completed projects. One recent report (Struttman and Holderman 1992) indicated costs of \$45,000 to \$53,000 for treatment of gasoline-contaminated soils over a 20 m (60 ft) radius to a depth of 6 to 10 m (20 to 30 ft) or a volume of 1,200 m^3 (1,600 yd^3). The system was designed with four recovery wells placed outside the "hot spots". The apparent air recovery rate of 4.8 m^3/min (170 ft^3/min) appears to be much higher than needed to provide oxygen for biodegradation. Offgas treatment was excluded under a permit exemption. The data from the site were not sufficient to determine how much gasoline removal was due to biodegradation or the extent of remediation that was achieved. It is not clear whether extended operations, which would lead to additional costs, were required.

As with many in situ remediation projects, economy of scale is an important consideration. In the above example, costs were approximately \$39/m^3

(\$30/yd^3). At another site, approximately 16,000 m^3 (21,000 yd^3) of soil were contaminated with an average of 5,000 ppm of diesel fuel (Downey, Hall, and Miller 1992). Soils consisting largely of silty sand, but interbedded with clayey soils, were contaminated to 21 m (70 ft). Initial pilot tests were conducted leading to an estimated cost of treatment of \$200,000, or \$13/m^3 (\$10/yd^3).

Summary. At this time, the basic principles of bioventing and understanding of what the gross factors that affect bioventing are well understood. As a result, good conceptual designs can be provided. Modeling of in situ vapor recovery provides a good basis for locating and designing air recovery wells. But, many critical questions remain, including the following:

- When is nutrient addition required? Does nitrogen fixation play a significant role?
- What is the relationship between respirometry results and actual bioventing treatment rates?
- What are the limits of concentrations of various types of contaminants that can be treated?
- To what levels can various contaminants be treated?
- What are the effects of cold weather?
- What is the effect of various venting rates on biodegradation?
- What are the limits of low-permeability soils?
- Can PAHs (or other compounds) be too tightly bound to soils to degrade?
- How easily can gas-phase constituents be reabsorbed and degraded?
- What are the optimum soil moisture concentrations in different soil types?

In addition, many questions about how system design and operating conditions affect the final outcome remain. For example, one can imagine a scenario where managing a site to maximize biodegradation over physical recovery can lead to nutrient deprivation that would not have occurred if the system were managed to increase the amount of physical removal. Given the documented differences between nutrient demands in laboratory tests and field tests, many questions will need to be answered in the field.

As with most remediation technologies, knowledge about bioventing is limited in part because much of the available information is too highly compartmentalized among a large number of organizations and individuals. This might be addressed by a clearinghouse to collect information in an appropriate form.

***3.5.1.1.3 Air sparging*[2]** In air sparging, air is introduced into the saturated zone to transfer volatiles to the unsaturated zone for biodegradation. Wells screened below the water table are used to inject air into the aquifer. As the air flows radially upward and outward in a series of air channels, volatile constituents are transferred to the unsaturated zone where they can be adsorbed by the soil and biodegraded by the indigenous microorganisms.

Air sparging in the unsaturated zone entails the use of high-air injection rates and/or closely-spaced injection wells to transfer volatile compounds to the unsaturated zone relatively faster than they are biodegraded in the saturated zone. By including the unsaturated zone in the treatment area, a larger volume of soil and potentially a greater number of microorganisms will be available for contaminant biodegradation than those contained in the saturated zone alone. At sites where the unsaturated zone consists of fairly permeable sands and is uncapped, it will be much easier to provide nutrients to the unsaturated than to the saturated zone.

It is necessary to control carefully the air injection flow rate to prevent transfer of volatile constituents to the atmosphere. In practice, it is likely that regulatory agencies or prudence will dictate some form of vapor recovery or other controls to prevent or limit losses to the atmosphere and, possibly, groundwater control as well.

Air sparging has been demonstrated at the U.S. EPA and Coast Guard demonstration project at Traverse City, Michigan, however, the data have not been reported yet. Air sparging in the unsaturated zone has not been documented or discussed in the cited reports on air sparging in the saturated zone. During sparging in the saturated zone, however, some portion of the

2. See Innovative Site Remediation Technology: Vacuum Vapor Extraction—Ed.

volatile components probably are transferred to the unsaturated zone and biodegraded.

3.5.1.2 Bioremediation Processes in the Saturated Zone

3.5.1.2.1 Liquid Delivery *Introduction.* R. L. Raymond and coworkers developed and patented the first bioremedial technique for treating aquifers contaminated with gasoline (Raymond 1974; Raymond, Jamison, and Hudson 1975; Raymond et al. 1978). The process is analogous to some conventional wastewater treatments in that a terminal electron acceptor, usually oxygen, and inorganic nutrients are added to enhance contaminant degradation in a bioreactor (Thomas and Ward 1989). But, unlike wastewater treatment, which takes place in a confined and easily-managed bioreactor, in situ treatment takes place in an unconfined and sometimes unpredictable bioreactor, the subsurface. Indigenous subsurface microorganisms are stimulated to degrade the contaminants.

Site Characterization. Before subsurface bioremediaton is initiated, extensive site characterization must be conducted (see Section 3.3 Site Characterization Relative to In Situ Bioremediation). The most important characteristic of a site is the presence of contaminant-degrading microorganisms. The relative number of these organisms will depend on the sediment characteristics, the type and concentration of contamination, and nutrients (Thomas and Ward 1992). Contaminant-degrading microorganisms have been detected in most samples of groundwater and sediments contaminated with petroleum compounds(see Subsection 3.2.1, Microbial Ecology and Physiology). Although increases in the number of hydrocarbon-degrading microorganisms can be detected in a matter of days after a petroleum spill, degraders of synthetic contaminants may require longer periods or may not develop at all. Low numbers of contaminant-degrading organisms may result from contaminant toxicity (Phelps et al. 1988) and low-nutrient availability; low numbers often are associated with sediments having high-clay content (nontransmissive) (Sinclair and Ghiorse 1989; Phelps et al. 1988).

To determine the feasibility of bioremediation, biodegradation (treatability) assays can be conducted in the laboratory using samples of contaminated sediment but not groundwater. The results of treatability tests using groundwater from unpurged wells may be misleading because groundwater collected from wells may contain non-native microorganisms

(Thomas, Lee, and Ward 1987). The potential for contaminant biodegradation and the nutrients required to maximize those rates are determined. Because of the highly carbonaceous nature of organic wastes, inorganic nutrient additions may be required. Studies are conducted to determine the types and concentrations of inorganic nutrients that maximize contaminant biodegradation rates; however, the native fertility of the subsurface should be considered before nutrient amendments are formulated. The macronutrients required are nitrogen and phosphorus, although additions of micronutrients, such as magnesium and manganese, may be helpful.

Although treatability assays provide information about biodegradation potential and nutrient amendments that enhance it, extrapolations from the laboratory to the field may not be exact, especially for rates. In situ biodegradation is limited by the rate at which oxygen is transferred to the contaminant-degrading microorgansims. In laboratory microcosms in which oxygen is not limiting, biodegradation is limited by the intrinsic metabolic rate of the microorganisms.

Computer models also have been used to assess the feasibility of bioremediation at sites contaminated with readily-biodegradable compounds, such as gasoline components (Norris 1993). In such instances, modeling the treatability of a site may be more cost-effective than conducting laboratory studies.

In addition to determining the nutrient requirements for the contaminant-degrading microorganisms, laboratory studies are conducted to determine the compatibility of the added electron acceptor and nutrients with the groundwater and subsurface materials. Nutrients that precipitate or complex with subsurface materials will not be bioavailable and may clog the aquifer (Thomas and Ward 1989). Well screens may be clogged by minerals that precipitate because the chemistry of the injection water and groundwater is different (Norris 1993). Injection of oxygenated water into aquifers containing low levels of dissolved oxygen may oxidize and precipitate iron and other metals. Orthophosphates, which are often added as sources of nutrients, will form calcium, magnesium, and iron precipitates which may clog the aquifer. Use of sodium phosphates can cause clays to swell and decrease the hydraulic conductivity of the formation.

Precautions can be taken to prevent nutrient incompatibility (Norris 1993). To avoid precipitation problems, water to be injected or re-injected can be filtered to remove metals and formation-compatable nutrients can be used. By using potassium, rather than sodium phosphates, in formations having high-clay content, problems associated with swelling clays can be avoided. Precipitation of calcium, magnesium, and iron with phosphates can be avoided through use of tripolyphosphate at equimolar or greater ratios, which will complex and solubilize these metals.

A thorough investigation of the hydrogeology and contaminant distribution is also required. Briefly, parameters such as hydraulic conductivity, depth to the water table, rate and direction of groundwater flow, specific yield of the aquifer, and types and concentrations of contaminants are determined. Analysis of sediments provides a better indication of the contaminant regime than does analysis of groundwater, since NAPLs and sorbed contaminants will not be detected in aqueous samples. Because the subsurface is heterogeneous, many parameters may be determined at multiple locations.

Contaminants sorbed to sediments and dissolved in groundwater are treated. For saturated sediments, physical removal of the contaminant source is more cost-effective and faster than biological treatment (Hurlburt 1987). Therefore, a free product phase should be removed before in situ bioremediation is implemented in saturated materials. Free product recovery, however, usually can remove less than 50% of the contamination in the subsurface.

System Design. Using information collected from the site characterization such as hydraulic conductivity, aquifer thickness, dispersivity, depth to the water table, and concentrations of contamination and oxygen, the well system can be designed with (Rifai et al. 1988) or without the use of computer models. Models also are used to predict the time required for remediation, based on the oxygen requirements for contaminant biodegradation and the rate of oxygen transport through the subsurface (Norris 1993).

The delivery system is designed to circulate adequate amounts of nutrients and oxygen through the zone of contamination to maximize contaminant biodegradation. Injection wells or trenches, through which nutrients and oxygen are added, are placed within or close to the contaminated area (figure 3.9 on page 3.95). Groundwater extraction wells or trenches may or may not be included, depending on the presence or absence of down-gradient receptors of concern. Produced groundwater is extracted, treated aboveground if necessary, and then disposed of in an environmentally-sound manner or amended with nutrients and recirculated. If the water is

Figure 3.9
Well System for Liquid Delivery

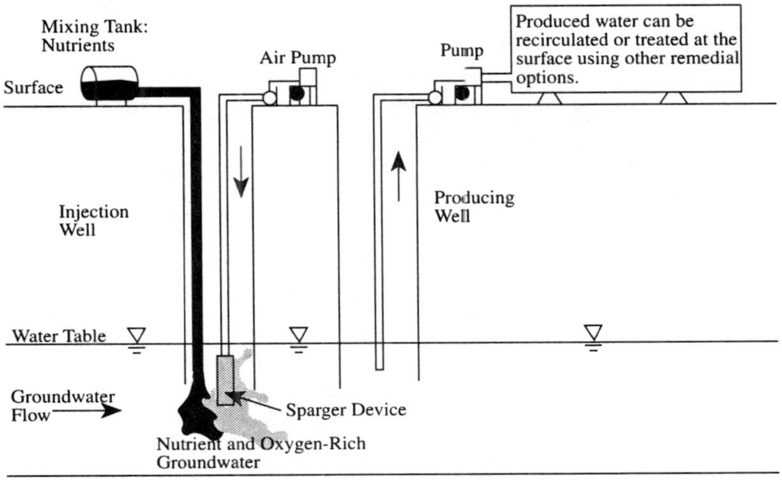

recirculated, removal of biomass and carbon may be required to avoid clogging the aquifer at the re-injection point. Many states require permits to inject any water into an aquifer, however, permits for injection of treated or clean water generally have not been difficult to obtain. Some agencies have limited the amount of nitrogen sources being injected into the aquifer to 10mg/L as nitrogen.

A recirculating system is designed to hydraulically isolate the target area and minimize contaminant migration out of the treatment zone. Some systems recirculate 100% of the produced water, whereas others use municipal water or uncontaminated groundwater from the site; in the latter case, the produced water must be disposed in an environmentally-sound manner. Other systems recirculate only a portion of the produced water to control groundwater flow through the isolated area when necessary. Where the unsaturated zone requires treatment or the aquifer is shallow, nutrients and oxygen may be added through infiltration galleries instead of injection wells (figure 3.10 on page 3.96). This variation allows the oxygen and nutrients to reach microorganisms in the unsaturated as well as in the saturated zone.

Figure 3.10
Infiltration Gallery for Treatment of Contamination in the Unsaturated Zone or Shallow, Contaminated Aquifers

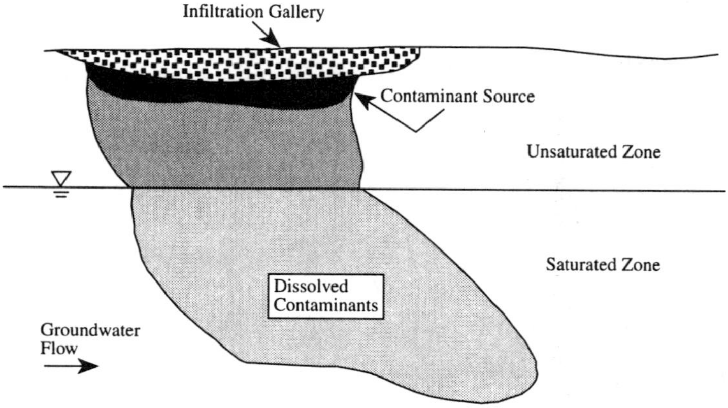

Monitoring wells must be installed in the treatment area and used to assess the progress of remediation. Samples of groundwater are collected before, during, and after bioremediation and analyzed for concentrations of oxygen, contaminants, nutrients, tracers, pH, and microorganisms. The monitoring program must be tailored to address several basic issues: regulatory requirements, monetary constraints, and hydrogeological aspects of the site. The first and second issues are self-explanatory. For the third issue, if groundwater flow rate is appreciable, it may be useful to have frequent monitoring early on to evaluate the possibility that the in situ system may produce quick results. On the other hand, when groundwater flow is slow, monitoring efforts can be spread out over time because it can be anticipated that not much will happen for a while.

Oxygen is added by sparging with air or pure oxygen at the bottom of an injection well, or by using hydrogen peroxide. This system can be modified so that air is added through spargers located directly in the aquifer (see Subsection 3.5.1.2.3, Air Sparging (Bioremedial Processes in the Saturated Zone)). The problem with using oxygen in an aqueous solution is its limited solubility in water. Depending on temperature, sparging with air or pure oxygen can yield about 8 and 40 mg dissolved O_2/L, respectively.

An alternate source of oxygen is hydrogen peroxide (H_2O_2), which is infinitely soluble in water and dissociates to form H_2O and 1/2 O_2. Although H_2O_2 can supply large quantities of O_2, the oxidant can be toxic to microorganisms at concentrations as low as 100 mg/L (Texas Research Institute 1982). To minimize toxicity, H_2O_2 is added initially at concentrations as low as 50 mg/L and increased step-wise to concentrations as high as 1,000 mg/L.

Although H_2O_2 potentially can deliver more oxygen than sparging with air or pure oxygen, the efficiency of delivering oxygen using the oxidant has varied (Norris 1993). If the rate of H_2O_2 decomposition is much greater than the rate of oxygen consumption, oxygen offgas may plug the aquifer and deplete the oxygen available to the microorganisms. In rare cases, the decomposition rate may be too slow and limit biodegradation. The rate of H_2O_2 decomposition is enhanced by the presence of organic matter, catalase (biological catalyst), and metals such as iron, copper, manganese, and chromium. In one field study, H_2O_2 was decomposed excessively by catalase-producing bacteria located in infiltration galleries (Spain et al. 1989). Decomposition rates can be enhanced by using catalysts such as chelated metals, or retarded by pretreating the aquifer with agents, such as inorganic phosphates that are also used as nutrients, that will inactivate the naturally-occurring catalysts (Raymond et al. 1986).

Although H_2O_2 can deliver more oxygen than sparging with air or pure oxygen, the ability to supply an electron acceptor for biodegradation using peroxide is still limited by the maximum theoretical solubility of oxygen in water (40 mg/L). An alternate electron acceptor that is more soluble in water than oxygen is nitrate (see Subsections 3.2.1, Microbial Ecology and Physiology; 3.2.2, Biogeochemistry and Biodegradation; and 3.5.1.2.2, Alternate Electron Acceptors); depending on the source, the solubility of nitrate is in the g to kg/L range. But, federal standards on the limits for nitrate in drinking water (10 mg/L as nitrogen and 45 mg/L as nitrate) will constrain the amount of nitrate that can be added to the subsurface.

Nitrate has been shown to serve as the electron acceptor for some aromatic hydrocarbons (Kuhn et al. 1988; Mihelcic and Luthy 1988). Nitrate has been used in field-scale trials in biorestoration of aquifers contaminated with petroleum hydrocarbons (Batterman 1986; Hutchins, Downs et al. 1991). (See also Subsection 3.5.1.2.2, Alternate Electron Acceptors).

Performance Monitoring. Monitoring is critical because of the dynamic nature of a bioremedial operation. Parameters that often are monitored include contaminant concentrations in groundwater and sediments, dissolved oxygen, inorganic nutrients, pH, and H_2O_2. Monitoring is important especially in the early phases of bioremediation to identify problems and the need to modify the system design and operation.

Interpretation of the monitoring results can be misleading if the parameters are analyzed singly rather than integratively. Analysis of groundwater for contaminant and nutrient concentrations is informative of progress, although it is not always indicative of removal of contaminants sorbed and entrained in the sediment. To accurately assess contaminant removal, sediment samples must be collected and analyzed periodically during treatment. During treatment, oxygen and/or H_2O_2 usually are not detected in samples of groundwater; the appearance of oxygen and/or H_2O_2 or an increase in their concentration (breakthrough) suggests that contaminants have been biodegraded in that area (Piotrowski et al. 1984). However, at sites with high contaminant concentrations and oxygen demand, rapid oxygen breakthrough may indicate flow through preferential flowpaths and non-uniform treatment.

Monitoring weekly changes in microbial numbers in groundwater as a measure of bioremedial progress can be misleading because of the dynamic nature of microbial communities in situ. Environmental parameters such as rainfall, predators, and microbial competition will affect population size. An initial decrease in microbial numbers after injection of water into the subsurface may be a result of dilution. But, a precipitous decline in microbial numbers and/or an increase in contaminant concentration may indicate inhibition of contaminant-degrading microorganisms.

Applications. The majority of sites that have been bioremediated using the liquid delivery technique have been contaminated with light nonaqueous phase liquids such as commercial blends of petroleum hydrocarbons. The lighter material (e.g., gasoline), containing the low-molecular weight, more soluble compounds, is more easily bioremediated than is the heavier material (e.g., coal tar), containing the high-molecular weight, less soluble compounds. Because high-molecular weight hydrocarbons are sparingly soluble and sorb to sediments, they are less bioavailable and more difficult to treat (Brubaker and Stroo 1992). Liquid delivery in the saturated zone has been demonstrated in the treatment of aviation fuel at Traverse City, Michigan (Wilson, Armstrong, and Rifai 1993). For dense nonaqueous phase liquids, contaminant distribution limits the applicability of liquid delivery. Unlike light nonaqueous phase liquids, the contamination is unevenly distributed rather than confined to a relatively small portion of the saturated zone that can be more easily treated.

In addition, the liquid delivery process functions best in formations with hydraulic conductivity of at least 10^{-4} cm/sec. For those formations with lower K values, transport of an electron acceptor and essential nutrients through the contaminated zone will require longer periods and be more costly. In addition, there is a greater chance of plugging the formation in the low-transmissive materials in which clays and fines often are found .

A variation of the process has been used to treat the chlorinated aliphatic solvents, such as trichloroethylene and less chlorinated ethylenes (Semprini et al. 1990). The process involves stimulating the growth of indigenous methanotrophs (methane oxidizers) with methane and oxygen. The methanotrophs metabolize methane using the enzyme, methane monooxygenase, which is nonspecific and also oxidizes the chlorinated ethenes.

Advantages/Disadvantages. All bioremediation processes have the potential to destroy the contamination rather than transfer it to another part of the environment; however, the most obvious and important advantage of the liquid delivery system is that treatment occurs in situ and obviates excavation and transportation of contaminated sediments. In contrast to pump-and-treat methods, which merely remove dissolved contaminants, in situ bioremediation processes can also treat contamination sorbed and entrained in the sediment. In addition, liquid delivery is faster than pump-and-treat.

Disadvantages are specific to the site. Bioremediation may be inhibited by the presence of toxic concentrations of contaminants. As previously discussed, formations with K values less than 10^{-4} cm/sec will require more time for treatment and are more easily plugged than those with greater K values. In addition, it may be difficult to obtain discharge permits to dispose of the produced groundwater that is not recirculated (Norris 1993).

Costs. Estimates of the cost for implementing liquid delivery will be specific to the site (Norris 1993). The type, amount, and extent of contamination will be important factors. In addition, the sediment characteristics will affect the system designed which must assure an adequate supply of oxygen and nutrients to the contaminant-degrading microorganisms. The source of oxygen will affect cost; although H_2O_2 is more expensive than oxygen, overall cost may be reduced when using the oxidant if the time required to treat the contaminants is reduced.

3.5.1.2.2 Alternate Electron Acceptors *Introduction.* Because of the limited solubility of oxygen in water, it is difficult to deliver large quantities of dissolved oxygen to contaminated subsurface environments. A variety of oxy-anions can subsitute for oxygen and allow microbial degradation of organic contaminants. Practical alternate acceptors include nitrate, sulfate, and salts of iron III.

Nitrate is highly soluble in water, has a high electron-accepting capacity for its mass, does not sorb appreciably to subsurface materials, and is not toxic to microorganisms. By using nitrate, large quantities of electron-accepting capacity are easily contained in groundwater circulated through a spill. But, nitrate is expensive and toxic to humans. Although nitrate is very soluble in water, the usual end product of nitrate reduction, N_2, is poorly water soluble. If N_2 accumulates, bubbles may form that exclude water from the pore spaces and decrease the hydraulic conductivity of the subsurface material.

Sulfate is also highly soluble in water, has a high electron-accepting capacity for its mass, and does not sorb appreciably. It is inexpensive and is not toxic to microorganisms. But, sulfide, the end product of sulfate reduction, is toxic to both humans and microorganisms.

Iron III salts are slightly soluble in water and have a low electron-accepting capacity expressed on a weight basis. It may be possible to distribute iron through contaminated material as a colloidal suspension in groundwater, but, this has not been demonstrated. More practical applications may involve mechanically blending iron minerals or iron salts with contaminated material. Iron II, the end product of iron reduction, is not particularly toxic at concentrations that would be expected during bioremediation.

Nitrate. In the early 1980s, Batterman (1986) used a mixture of nitrate and oxygen to remediate a large spill of heating oil in the upper Rhine Valley. Benzene and toluene were completely removed, but the xylenes were more persistent. This remains the most successful large-scale application of an alternate electron acceptor for remediation of a fuel spill. Hutchins,

Downs et al. (1991) conducted a detailed study of the use of nitrate to remediate a JP-4 jet fuel spill in Michigan. Benzene, toluene, ethylbenzene, and all three xylenes were removed from groundwater circulated through the spill. Core analyses also revealed that these compounds were removed from the residual oily phase hydrocarbons left behind after active remediation ceased. Microcosm studies conducted with core material from the site confirmed removal of toluene, ethylbenzene, *m*- and *p*-xylene under strict denitrifying conditions (Hutchins, Sewell et al. 1991). *o*-Xylene, however, only degraded when one of the other alkylbenzenes was present. Benzene was not degraded at all, whether present as the sole carbon source, or in combination with the alkylbenzenes.

Removal of benzene in the field may have resulted from low concentrations of oxygen (0.5 to 1.5 mg/L) in the groundwater circulated through the spill (Hutchins, Downs et al. 1991). But, the precise role of oxygen has not been defined. The alkylbenzenes can be degraded under strictly anoxic conditions through pathways involving oxidation of the alkyl substituent. These pathways are not available to benzene. Oxygen may be required for oxygenases that initiate the metabolism of benzene.

Lemon, Barbaro, and Barker (1988) studied the efficacy of nitrate for bioreclamation of an artificial plume of alkylbenzenes in Ontario, Canada. Under strictly denitrifying conditions, only toluene was removed. The aquifer had previously been aerobic and there was little time for anaerobic acclimation to occur at the time the experiment was conducted.

Hutchins and Wilson (1991) noted that removal of alkylbenzenes in the JP-4 jet fuel spill was a zero-order process. At 10°C (50°F), toluene was removed at a rate of 0.2 mg/L/day, ethylbenzene at 0.13 mg/L/day, *m*- or *p*-xylene at 0.14 mg/L day, *o*-xylene at 0.13 mg/L/day, and 1,2,4-trimethylbenzene at 0.073 mg/L/day.

Nitrate also has been used to reclaim a gasoline spill in California (Sheehan et al. 1988). From 95% to 98% of the purgeable alkylbenzenes in groundwater were removed.

Sulfate and Iron III. As of this writing, neither sulfate nor iron III has been used as an electron acceptor in a field-scale remediation, although both sulfate and iron III have important roles in natural bioattenuation. Recent work in Australia demonstrated that ambient concentrations of sulfate (20 to 100 mg/L) were responsible for removal of toluene, ethylbenzene, the xylenes, and 1,3,5 trimethylbenzene from a plume of contamination from a leaking underground storage tank (Thierrin et al. 1992a). Benzene was not removed. Lovley and Phillips (1988) have reviewed the importance of iron reduction in the natural carbon cycle. An organism has been isolated that degrades toluene, phenol, and *p*-cresol with amorphous iron III as sole electron acceptor (Lovley and Lonergan 1990).

Sulfate is much less expensive than nitrate. Concerns about sulfide toxicity are probably responsible for the lack of interest in sulfate as an alternate electron acceptor. Recent laboratory work suggests a practical solution to the sulfide toxicity problem. Beller, Grbic-Galic, and Reinhard (1992) studied microcosms and enrichment cultures prepared from soil that had been contaminated with fuel hydocarbons for a long period. Sulfate reduction supported removal of toluene in the microcosms and enrichment cultures. Addition of iron III significantly stimulated the removal of toluene by sulfate reduction. The degradation of toluene, reduction of iron III, and removal of sulfate were consistent with formation of elemental sulfur, iron II sulfide, and iron III carbonate, but there was no free sulfide. If iron minerals are a significant component of the aquifer matrix, it may be possible to use sulfate effectively as an alternate electron acceptor.

The Question of Benzene. Complete mineralization of benzene under anaerobic conditions was recently documented in samples of sediment from Seal Beach, California (Edwards and Grbic-Galic 1992). The electron acceptor was not confirmed, but was probably sulfate. The experimental system contained 1,900 mg/L sulfate, no oxygen or nitrate. Benzene was not degraded in material that also contained other alkylbenzenes. Benzene degradation in the absence of oxygen is possible, but it is also rare and unpredictable at the present time. In the few cases where it can be inferred from field data, and the one case where it is confirmed with a laboratory study, it was associated with sites that have been exposed to petroleum hydrocarbons for years or decades. For the present, engineering designs to bioremediate benzene in situ should consider adding oxygen in addition to the alternate electron acceptor.

3.5.1.2.3 Air Sparging *Introduction.* Air sparging provides oxygen as an electron acceptor for biodegradation and physically removes volatile substances from the unsaturated zone. Air is forced into the aquifer through well points or wells screened beneath the water table. Air moves radially outward and upward from the point of injection, resulting in increased levels of dissolved oxygen and transfer of volatiles to the unsaturated zone. Dissolved oxygen is distributed through the aquifer through diffusion and the movement of air, and groundwater movement. Volatiles reaching the unsaturated zone are typically captured using in situ vapor recovery with offgas treatment. A recent field study has demonstrated the use of the unsaturated zone as a biofilter, thus avoiding the need for expensive offgas treatment as discussed elsewhere in this monograph.

The way in which air moves through saturated soils is not well understood. Some authors, in early publications, described air flow during sparging as a series of bubbles. Currently, it is believed that bubbles may only be present in significant quantities in coarse gravels. It is more likely that in most soils, air will flow through discreet channels. The spacing between discrete continuous flow channels is not well understood, but is critical to the effectiveness of air sparging, especially for physical removal of volatile constituents.

Therefore, air sparging can be thought of as functioning through one or more of the three following mechanisms:

- biodegradation in the aquifer;
- biodegradation in the unsaturated zone; and
- physical transfer to the unsaturated zone for capture by an in situ vapor recovery system.

The relative importance of the mechanisms will depend on the properties of the contaminants, the site conditions, and the system design. This section discusses systems which emphasize biooxidation in the aquifer. Systems designed to emphasize biooxidation in the unsaturated zone are discussed elsewhere in this monograph. A comprehensive discussion of air sparging as a means of promoting physical transfer can be found in the companion monograph on in situ vapor recovery.[3]

The use of air sparging for biodegradation in the saturated zone has evolved through efforts to improve upon the liquid delivery system. The liquid delivery system supplies oxygen by sparging air or pure oxygen at the bottom of injection wells. Because the low solubility of air in water limited the rate of oxygen delivery, hydrogen peroxide, which is infinitely

3. See Innovative Site Remediation Technology: Vacuum Vapor Extraction—Ed.

soluble in water, was used as an alternate source of oxygen. Practical considerations have limited hydrogen peroxide concentrations to the 100 to 1,000 mg/L range. Using hydrogen peroxide, greater rates of oxygen delivery were achieved than were possible by sparging into the injection water and have led to successful bioremediation of aquifers. However, there are problems with the use of hydrogen peroxide other than its relatively high cost (see section on Liquid Delivery in the Saturated Zone) (Norris 1993; Lowes 1991). Sparging air directly into the aquifer offers a potentially technically preferable and more cost-effective method of oxygen introduction than the liquid delivery system.

A much larger volume of oxygen can be introduced into the aquifer by air sparging than by other technologies. Very close to the air channels, it may be possible to maintain dissolved oxygen near saturation. However, it is not yet understood how much dissolved oxygen is maintained at points midway between air channels. Further, there is currently insufficient information to predict the fraction of the oxygen introduced below the water table that is transferred into the aqueous or dissolved phase. Bioremediation systems using air sparging would use many more and more closely-spaced air sparging wells compared to the number and spacing of injection wells used for a hydrogen peroxide system. The belief of many practitioners is that this will result in a greater oxygen availability across the contaminated zone.

Furthermore, direct air sparging introduces oxygen across the entire contaminated zone during the treatment period, rather than relying on oxygen transport through the contaminated zone. Thus, air sparging will more likely result in remediation of all portions of the site at similar rates, as opposed to a degradation front moving across the site during liquid delivery.

System Design. Sparging wells are placed at short intervals located near the bottom of the aquifer (figure 3.11 on page 3.105). Air is supplied at a sufficient pressure to overcome the head pressure of groundwater and surface tension. The outward and upward movement of air through the saturated soil matrix has the following effects:

- as the air bubbles move through the groundwater, the elevated pressure of the air bubbles induces air and, therefore, oxygen to be dissolved into the aqueous phase. Diffusion accelerates the

Figure 3.11
Groundwater Sparging Without Optional Vapor Recovery

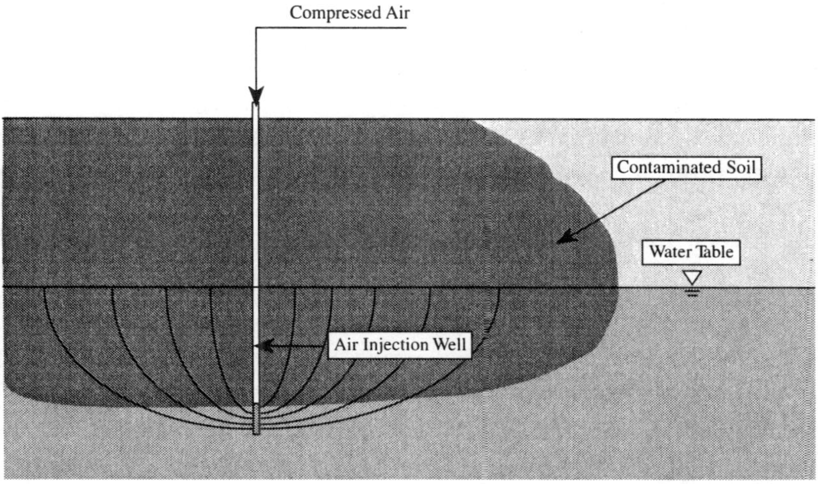

movement of oxygen and mitigates the effects of channeling of the air flow;

- while air moving through the soil pore spaces creates turbulence, improving soil and water contact and, thereby, increasing the rate of dissolution of the contaminants, the greatest amount of mixing is thought to occur during the transient stage following startup and shut-down;
- the movement of air through the groundwater acts as a crude air stripper, transferring dissolved compounds with sufficiently large Henry's Law constants to the unsaturated zone;
- adsorbed-phase volatile compounds can be transferred to the air stream and carried to the unsaturated zone;
- the temporary mound observed in air sparging is caused by pressure and the displacement of water from pores that become air filled; and

- physical displacement of free-phase substances, especially dense nonaqueous phase liquids (DNAPLs), may occur at very high air flow rates.

The degree to which biodegradation will occur (versus physical transfer to the unsaturated zone) will depend on the characteristics of the contaminant(s), the site lithology, and the design of the system. For heavier hydrocarbon blends, volatilization will be minimal regardless of the system design and site conditions. For the lighter petroleum hydrocarbons, both biodegradation and physical removal processes should be considered in the system design and operating conditions. In this section, systems whose objective is to maximize contaminant reduction via biodegradation in the saturated zone will be discussed.

The design and implementation of air sparging systems for bioremediation requires a basic understanding of biodegradation, volatilization of organic compounds, and air movement; however, the effects of air movement through saturated soils is currently not well understood. Although biodegradation is the goal of the process, it also is necessary to control volatile organic compounds that reach the unsaturated zone through in situ vapor recovery systems (figure 3.12 on page 3.107). Because of the heterogeneous soil conditions in most aquifers, it is typically necessary to conduct field tests and use other site information to engineer a safe and reliable remediation system.

Bacteria require nutrients, especially nitrogen and phosphorus, to grow and thereby degrade organic contaminants. Accordingly, it is generally accepted that nutrient addition is necessary for most bioremediation processes. But, the results of recent bioventing field tests have raised questions concerning the necessity and benefits of adding nutrients for bioremediation of the unsaturated zone (Downey and Guest 1991; DuPont 1992a; Miller 1990; Hinchee et al. 1991). Nutrient addition may not be necessary in unsaturated zone treatment. It has been speculated that nutrient addition is more important for bioremediation of the heavier, more recalcitrant compounds on which bacteria grow slowly, than for remediation of the more readily-degradable substances on which bacteria grow faster (McKenna and Heath 1976). At this time, there is not enough information to determine whether nutrient addition is necessary. Until there is better understanding, conceptual designs of these systems should provide for nutrient addition. The decision to employ nutrient addition can be decided during pilot testing or implementation.

Because addition of air to the saturated zone displaces water, it is common to use a groundwater capture system to prevent migration of dissolved contaminants. A portion of this captured water can be used to inject dissolved nutrients into the aquifer. Another benefit of groundwater recovery and re-injection is the movement of water between air-flow channels, thus increasing the movement of dissolved oxygen containing water. In shallow aquifers, nutrients can be percolated from the surface. A large body of experience exists on distribution of nutrients into aquifers for bioremediation exists. There have been few concerted efforts, however, to select nutrient blends for specific conditions. See also Subsection 3.5.1.1.2, Bioventing, concerning nutrient addition.

While air sparging has been used for the physical removal of volatiles from the unsaturated zone, as discussed by Brown and Jasinlewicz (1992) and Brown and Fraxedas (1992), the movement of air through saturated soils is far from fully understood. Several field tests have been performed that established radii of influence under specific site conditions. One labo-

Figure 3.12

Groundwater Sparging With Optional In Situ Vapor Stripping for Management of Vapors

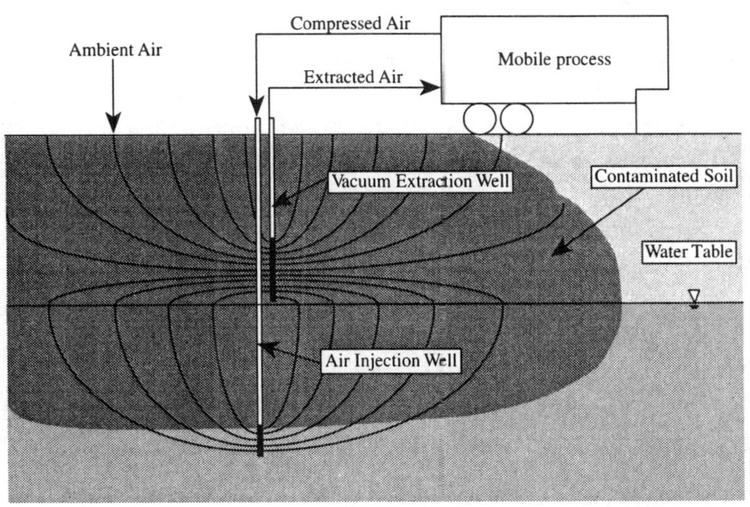

ratory study that was conducted to determine how the radius is affected by flow rates (Wilson, Clarke, and Clarke 1992) indicated that the radius of influence was approximately equal to the saturated thickness; channeling within the test apparatus soils was also observed. Other laboratory studies also demonstrated the formation of channels and indicated that the radius of influence increased with increased heterogeneity of soil particles. One recent paper developed a mathematical model for correlating radii of influence with the distribution of elevated pressure in the vadose zone (Wilson, Clarke, and Clarke 1988). Rough rules for estimating radii of influence of sparging wells predict that the air bubbles will reach the groundwater surface at a horizontal distance from the injection point equal to the depth below the water table at which the air is introduced. This assumes that there are no low-permeability lenses that impede upward movement of the air and does not consider diffusion of oxygen, which is important for biodegradation but not for physical removal of volatile compounds.

Despite the relatively undeveloped status of the use of air sparging for bioremediation, the information and experiences derived from air sparging for physical removal of volatile compounds and experiences with more traditional aquifer bioremediation can be used to intelligently implement sparging to treat aquifers. As with all in situ technologies, it is necessary to have a good understanding of the environmental conditions and the properties of the constituent contaminants. It is advisable, however, to carefully monitor performance to verify that the intended results are occurring or to identify corrective measures that can be taken.

The mechanism for treating compounds in the saturated zone under air sparging conditions will depend on the relative ease of biodegrading the compounds compared to the ease of stripping them from the aquifer. The latter will depend on the compounds' water solubility, vapor pressure, and physical state within the aquifer (dissolved, free phase, or adsorbed). Dissolved compounds will be most easily stripped if they have a low-water solubility and a high-vapor pressure. The tendency to be stripped from the dissolved phase is most directly related to a compound's Henry's Law constant (table 3.18 on page 3.109). For example, acetone, which has a higher vapor pressure.than benzene or TCE, has a significantly lower Henry's Law constant because it is more soluble in water and thus less easily lost to the vapor phase. Compounds with a Henry's Law constant greater than 1×10^{-3} m^3/mole will have a significant tendency to transfer to the vapor phase. Physical properties that will affect the disposition of many compounds of

Table 3.18
Physical Properties Important to Bioremediation Processes

Compound	Solubility (mg/L)	Log K_{ow}	Vapor Pressure (mm Hg)	Henry's Law Constant (m^3/mole)
Benzene	1,791	2.13	95	5.4×10^{-3}
TCE	1,100	2.40	69	1.0×10^{-2}
Acetone	Misc.	-0.24	231	3.7×10^{-5}
Butanol	77,000	0.88	7	5.6×10^{-6}
Phenol	87,000	1.46	0.5	4×10^{-7}
Naphthalene	31	3.28	0.2	4.6×10^{-4}
Pyrene	0.1	5.20	6.9×10^{-7}	1.1×10^{-5}
Diethylhexylphthalate	0.3	5.11	5.0×10^{-6}	1.1×10^{-5}
Phenanthrene	0.9	4.46	6.8×10^{-4}	3.9×10^{-5}

From Howard 1989, 1990; Montgomery and Wilkins 1990

environmental concern can be found in several handbooks (Howard 1989, 1990; Montgomery and Wilkins 1990; Verschueren 1983).

Many compounds with relatively high Henry's Law constants, such as benzene, are also readily biodegradable under aerobic conditions. Compounds with low Henry's Law constants, such as phenol, will readily undergo biodegradation, but will not be easily removed through volatilization. Thus, there is more flexibility in bioremediating the former compounds with relatively high Henry's Law constants, for which designs can take advantage of their volatility and ease of biodegradation.

Following are the site characteristics that are important for air sparging:

- contaminant identification, including properties;
- contaminant levels and distribution in each phase;
- depth to groundwater;
- thickness of aquifer;
- soil types and distribution;
- surface conditions;
- utility trenches;

- presence of receptors; and
- groundwater flow rate and direction.

Identification of the contaminants is important so that the properties of the compounds can be evaluated for sparging suitability. The contaminant levels and distribution are needed for predicting oxygen and, possibly, nutrient requirements and evaluating potential for inhibition.

The hydrogeological information is important because the site hydrogeology will determine whether air can be injected at sufficient rates, the flow patterns that will result, and the design of the air injection system.

Air will obviously be more readily introduced into coarse than fine soils. Air injected into soils will tend to move through the more permeable zones. This can lead to channeling and result in areas of the contaminated zone that are not treated. If air is injected below a confining layer or clay lens, the air will move laterally before reaching the water table.

Figures 3.13a to 3.13c show the effect of site hydrogeology on air-flow patterns and, therefore, on the systems' design and performance. In figure 3.13a, the soils are relatively homogeneous. Air travels radially from the sparging well and reaches the vadose zone at a distance approximately equal to that from the water table to the top of the sparging well screen. In figure 3.13b (on page 3.111), the soils are relatively fine, except for a coarse sand layer located near the bottom of the saturate interval. In this type of matrix, the air flows preferentially through the coarse layer, signifi-

Figure 3.13a
Air Sparging in Homogeneous Aquifers

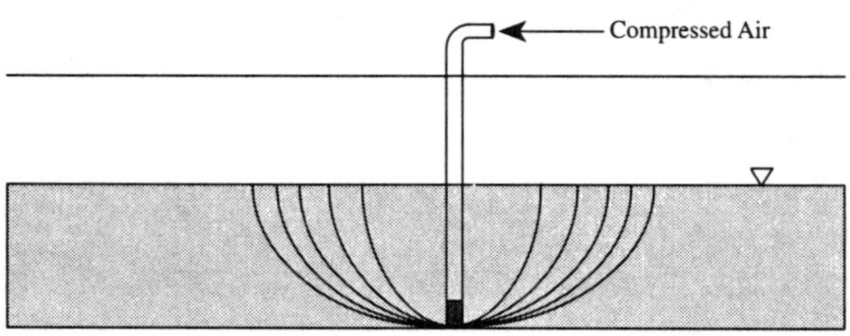

Figure 3.13b
Air Sparging in Aquifers With Gravel/Sand Layer

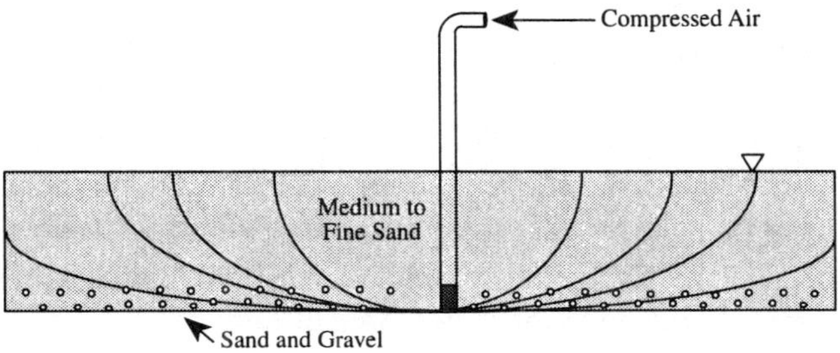

cantly increasing the lateral movement of air. But, the injected air may not pass through soils above this layer. Similarly, in figure 3.13c, less permeable soils near the top of the aquifer cause lateral movement of the air before it can reach the vadose zone.

The effect on oxygen distribution attributed to stratified layers in the saturated zone was observed during a treatability test of an air sparging system. A 0.3 m (1 ft) thick gravel and sand layer near the bottom of a 2 m (6 ft) thick saturated interval consisting of silt with some sand resulted in dissolved oxygen levels increasing from 1.2 mg/L to 5.5 mg/L at a distance

Figure 3.13c
Air Sparging in the Presence of Clay Lenses in Upper Portion of Aquifer

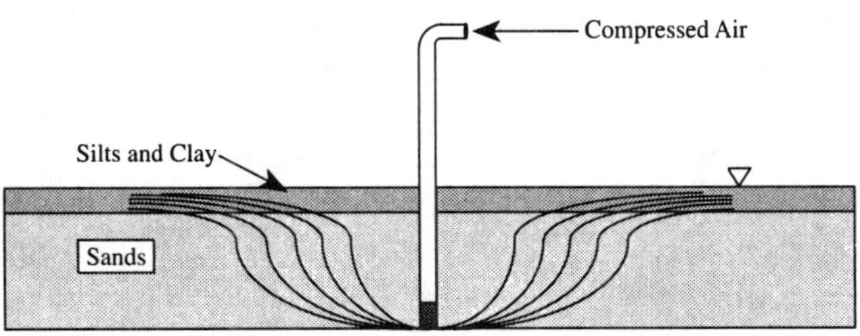

of 10 m (32 ft) in less than 2 hours (Norris and Clarke 1991). Dissolved oxygen levels were increased at a distance equal to at least five times the saturated interval above the sparging well screen.

Depth to groundwater is important in treating aquifers where volatile components are present. Regardless of the design and operating parameters, some volatile compounds will be carried into the unsaturated zone. When the water is very shallow, it is difficult to design an efficient vapor recovery system unless the surface is capped. Furthermore, the period during which contaminants travel through and can be biodegraded in a thin vadose zone is short. On the other hand, shallow water tables may allow nutrients to be introduced to the water table through percolation.

For air sparging, the radius of influence is typically defined as the greatest distance at which the air reaches the upper surface of the aquifer. When air sparging is used to promote biodegradation, the radius of influence of interest is the greatest distance at which significant increases in dissolved oxygen can be attained. The radius of influence for air sparging is a useful quantity in that it is the minimum radius of influence for sparging to promote biodegradation.

A conceptual design for air sparging can be developed by assuming that the radius of influence of air reaching the water surface is approximately equal to the saturated interval to a depth of 9 m (30 ft). At greater depths, uncertainties concerning air flow increase. The radius over which dissolved oxygen levels can be increased can be significantly larger than the radius for physical removal. This information can provide an indication of feasibility. But, since soils are typically heterogeneous and there has been limited documented experience with air sparging systems, it is necessary to conduct field studies to determine the radius of influence at each site.

For aquifers contaminated with nonvolatile constituents only, it is sufficient to determine the radius over which increased dissolved oxygen will be attained. For several reasons, dissolved oxygen measurements should be used as semiquantitative data. This can be accomplished using a sparging well and groundwater wells or drive points. The sparging well should be installed close to the bottom of the aquifer. Continuous split-spoon samples should be taken and carefully logged, unless nearby boreholes or wells have already been sampled in this manner. Wells or drive points should be installed along a straight line at distances equal to one-half, one, two, four, and eight times the distance between the top of the air sparging well screen and the water table. At least one additional well should be installed along a line perpendicular to the line of monitoring points. Air should be injected into the sparging well sequentially at three increasing pressures and flow rates. The dissolved oxygen levels should be determined at decreasing time intervals after each change in operating conditions. Sparging rates can be increased without interruption of the air flow; however, air flow can be terminated for several hours before testing a higher rate of air flow to allow air to be dispersed uniformly throughout the treatment zone. The latter method is preferred if channeling is likely to be a problem.

In aquifers where volatile compounds are present, it is also necessary to monitor the unsaturated zone for radius of influence and for these compounds. In addition to one or more sparging wells and several monitoring points in the saturated zone, monitoring points in the unsaturated zone and. typically, a vapor recovery system will be required. Monitoring points in the unsaturated zone can be wells screened across both the saturated and unsaturated zone, or separate monitoring points in the unsaturated zone. In shallow aquifers, soil-gas monitoring probes can be used to sample the unsaturated zone. The low cost and ease in installing these probes allows extra soil-gas sampling points to be inserted as the test proceeds to more precisely locate the radius of influence.

The radius of influence has been estimated by measuring increases in air pressure at probes located above the water table. The drawback of this approach is that increases in air pressure will extend past the point where air reaches the water table. A paper by Wilson et al. (1992) describes a mathematical model that uses air pressure readings from several probes to estimate the actual radius of influence. Another approach is to add a tracer gas, such as helium or sulfur hexafluoride, to the sparged air and monitor the discharge of an operating vapor recovery well located proximate to the air injection well; the vapor recovery well must be operated at a flow rate exceeding that of the sparging well (Norris and Mutch 1991). Air reaching the unsaturated zone will contain low levels of the tracer gas. The gas that reaches the unsaturated zone will be captured by the vapor recovery well. Thus, any monitoring point from which tracer gas is detected must be within the region in which the air bubbles reached the water table.

For very shallow aquifers, it may be possible to observe air in vertically inserted pipes that extend a few centimeters into the water table. This method gives the most direct indication of air reaching the water table and, because of the ease of installation, several points can be installed before or during the field test.

Some test protocols call for measuring the radius of influence sequentially during air sparging only, air recovery only, or air sparging and air recovery together (Brown 1993). Radii of influence induced by both air sparging and vapor recovery and the concentrations of volatile constituents in the recovered air are monitored. This approach provides design data for placing air recovery and air sparging wells and an indication of the transfer of volatile constituents to the unsaturated zone.

Several design strategies are possible. Systems can be designed to minimize remediation times by maximizing both biodegradation and physical removal at a cost of increased capital and operating costs (especially for offgas treatment). Other systems can be designed to minimize costs by favoring biodegradation over physical removal, which is likely to require longer to reach clean-up goals.

In one design, sparging wells are placed in staggered rows (figure 3.14), commonly referred to as a "five spot." Staggering adjacent rows improves the overlap between wells. The wells should be spaced at 70 to 80% of the radius of influence to provide complete overlap. To increase the probability of all areas of the site being remediated in the same time frame, spacing should be closer in the most highly-contaminated areas.

In formations free of rubble and large stones, it may be possible to use one or more horizontally installed wells. This approach was used at the Savannah River Project for stripping chlorinated solvents from the saturated

Figure 3.14
Plan View of Staggered Rows of Air Sparging Well

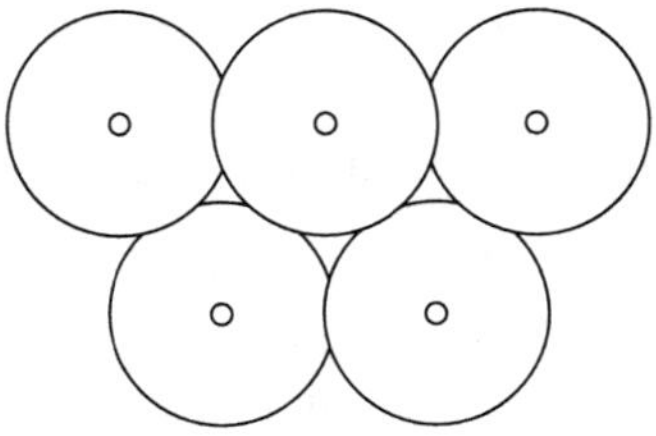

zone and in an ongoing test to inject air and methane to promote cometabolism of the residual chlorinated solvents (Schroeder et al. 1992). The design called for the addition of methane, as well as air and nutrients, to the aquifer to promote degradation of chlorinated solvents through cometabolism.

For sites with volatile constituents, it is likely that an air recovery system will be needed unless biodegradation can be completed in the unsaturated zone. The same type of air recovery systems as discussed in Subsection 3.5.1.1.2, Bioventing, can be used with the same strategies for either physical removal or biodegradation as the primary mechanism for removal of volatile compounds from the unsaturated zone.

In most instances, the radius of influence of air recovery wells will be greater than that of sparging wells, particularly for capped sites. Thus, fewer air recovery wells than air sparging wells will be normally incorporated into the design. The total flow of the air recovery wells, however, must be greater than the total flow of the air sparging wells. Spacing and location of recovery wells relative to air sparging wells will depend on the concentration of volatile compounds and whether the intent is to enhance biodegradation in the unsaturated zone or treat the contaminants at the surface.

In many instances, it may be necessary to capture groundwater to prevent migration of dissolved constituents during remediation. Designs for typical pump-and-treat systems should suffice, taking into account that the air sparging system may, in some geological conditions, increase the lateral movement of the groundwater.

Nutrient addition may or may not be needed and feasible. If nutrient addition is required and is very difficult or infeasible, it will probably be best to design and operate the system to promote physical removal of volatile compounds rather than biodegradation.

For shallow aquifers, particularly those with sandy soils, nutrients may be most effectively added by percolating from the surface. In general, recovered groundwater can be amended with nutrients and then re-injected.

System Operation. Operation of air sparging systems to maximize the biodegradation contribution requires that adequate nutrient levels are present while initially injecting oxygen for a short duration with relatively long times between air injection periods. This minimizes air stripping during a lag period and allows sparging wells to be operated intermittently. To be efficient, the wells must be operated at sufficient pressure to have adequate radii of influence. Air is then introduced at much faster rates than can be used for biodegradation. Thus, it is only necessary to operate the wells for short periods (e.g., 0.5 to 1 hr) between relatively long inoperative periods (e.g., 12 to 48 hr). The off/on schedule can be developed based on dissolved oxygen levels in groundwater and volatile compounds in the unsaturated zone.

A fortuitous effect of the off/on operating schedule is the change in the groundwater elevations (Brown 1993). During sparging, the groundwater level rises an increment in the vicinity of the sparging well, in addition to the rise resulting from air extraction. After a short period of operation or when sparging is interrupted, the water table drops below the equilibrium level that existed before sparging and then rises to the equilibrium level. The soils near the water table in the vicinity of the sparging well alternate between being saturated and unsaturated; thus this interval may experience the greatest benefit as a result of increased oxygen flow through unsaturated as well as saturated soil and maintaining high levels of moisture. As a result, these soils are subject to the highest levels of physical removal.

During operations, measurements should include:

- concentrations of carbon dioxide, oxygen, and volatile compounds in the vadose zone;
- contaminant concentrations, dissolved oxygen (and maybe carbon dioxide), and pH in groundwater;
- groundwater levels during and between air injection; and
- flow rates and air pressure of the air sparging and vapor recovery systems.

Insufficient information has been reported about air sparging to establish rules of good practice. But, it is essential that persons involved in the design and implementation of this technology understand the principles involved and follow a generally conservative approach, especially with respect to the potential for uncontrolled migration of volatile compounds within the unsaturated zone and for migration of dissolved constituents. Some level of pilot testing is required in developing remedial designs, and designs should be flexible so that additional components can be added to the system.

This technology does not present any unique health and safety requirements other than that of insuring that vapor and dissolved phase migration are monitored and controlled, if necessary. However, great care does need to be taken with regard to migration of vapors into buildings and utility trenches. Otherwise, following the regulations prescribed under the Occupational Safety and Health Act (OSHA) should be adequate.

3.5.1.3 Costs of In Situ Bioremediation Technologies

The cost of in situ bioremediation is site- and contaminant-specific. The types and concentrations of contaminants, the volume of contaminated material, permeability, soil characteristics, clean-up standards, depth to water table, monitoring requirements, and site location will affect cost (Norris, 1993). Other factors that are specific to the technology will also affect cost (Table 3.19).

Table 3.19
Costs of In Situ Bioremediation Technologies

Technology	Waste	Technology-Specific Factors Affecting Cost	Cost ($/m^3)
Land treatment	Petroleum hydrocarbons; contaminated soil; sludge	liner; runoff requirement	$10 - 80
Bioventing	Petroleum hydrocarbons	- offgas treatment	$10 - 20
		+ offgas treatment	$52 - 78
Liquid delivery	Petroleum hydrocarbons	oxygen source	$50 - 200
Air sparging	Petroleum hydrocarbons		$25 - 150

US EPA 1991

3.5.2 Ex Situ Bioremediation Technologies

Ex situ bioremediation technologies are those in which a waste that has been removed from its point of origin is treated in a closed or open bioreactor. The types of wastes that can be treated include liquids, solids, and air. Although a bioreactor is commonly thought of as a vessel in which the waste is treated in a controlled manner, there are other forms, including lined lagoons, soil-piles, composting piles, and soil (bio) filters. All bioreactors, however, are designed and managed to maximize the biological reaction in an economical manner. The reactor design solves particular problems that are encountered with the contaminated wastes under consideration.

There are two basic problems to be solved in any design of an aerobic bioreactor. The first is that of contact between the bacteria and the organic contaminants. The second is that of oxygen transfer to the bacteria. The various bioreactor designs that are available can be compared by the way they solve these two problems. Other criteria also will be used to establish the advantages and disadvantages of specific reactor designs, but bacterial contact and oxygen transfer are the two functions that are common to all reactor designs.

Bacterial contact amounts to more than merely mixing the bacteria with the organic contaminants. The goal of the biological reaction is to destroy a maximum amount of the contaminants and to leave a minimum concentration of the contaminants. To achieve these goals, the bacteria must be put in contact with the contaminants and given extended periods to complete the biochemical reactions. In other words, the bacteria must have a long residence time in the reactor. Figure 3.15 (on page 3.119) shows the relationship between the effluent organic concentration and the residence time of the bacteria.

Oxygen transfer does not affect the performance of the reactor design as long as a minimum oxygen concentration is maintained. Oxygen transfer is related mainly to the cost of biological treatment. Energy for oxygen transfer is usually a main operating cost of a bioreactor, exceeded only by labor.

3.5.2.1 Ex Situ Treatment of Contaminated Water[4]

Many designs for biological treatment of contaminated groundwater are based on systems originally designed for treating relatively high organic strength wastewater. It is inappropriate to assume that if "activated sludge" cannot be used successfully to treat the lower organic loads associated with groundwater contamination problems then biological treatment will not

4. Reprinted by permission of Van Nostrand Reinhold Publishers from "Groundwater Treatment Technology," Second edition, by E. Nyer. Copyright 1992 by Van Nostrand Reinhold Publishers.

Figure 3.15
Effluent Organic Concentrations With Increasing Bacterial Residence Time

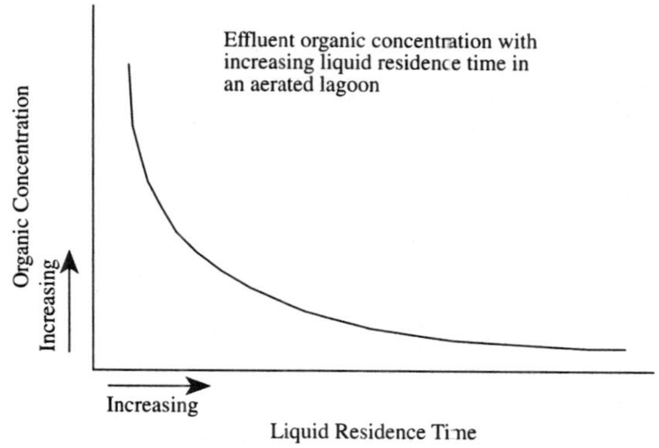

work. If a biological reaction is possible, then biological reactors can be designed in such a way that problems specific to groundwater treatment are addressed. Several designs have recently been developed specifically for low concentrations. Similar problems must be overcome when treating liquid waste streams emanating from pond remediations.

Contaminant concentrations can follow three patterns over the life of a project (figure 3.16 on page 3.120). First, there is the constant concentration exhibited by a leachate. If the source of contamination is not removed, the source will replace the contaminants as fast as they can be removed with the groundwater system, and the concentration will remain constant.

The second possible pattern arises when the contamination plume is being drawn toward a groundwater removal system, such as a municipal drinking water well system. In this situation, contaminant concentrations increase over time. The well is originally clean, but becomes more contaminated as the plume is drawn toward the well.

The final pattern is that incident to remediation. If the original source of contamination is removed, the concentration of the contaminants will decrease over time. The reduction in contaminant concentration is a result of

Figure 3.16
Time Effect on Concentration Found in a Well

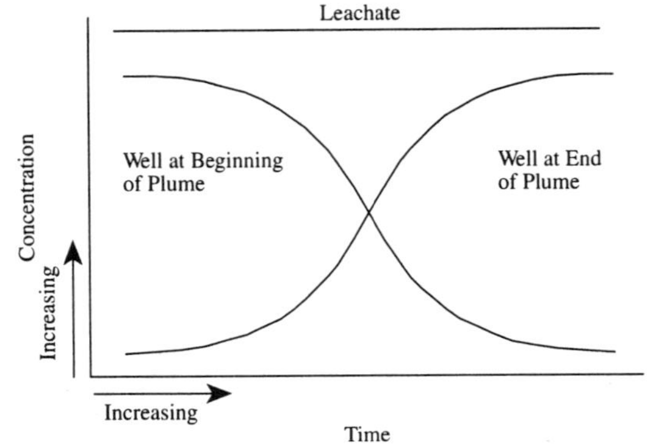

retardation, natural chemical and biochemical reactions, and dilution by the surrounding groundwater.

A bioreactor designed for groundwater must be able to treat the contaminants over the entire life cycle of the remediation. Simply designing the system to treat the initial concentration is not sound engineering design.

Bioreactors that treat contaminated water can be separated into five main groups: suspended-growth reactors, fixed-film reactors, reactors based on activated carbon, submerged fixed-film systems, and miscellaneous designs. In suspended-growth reactors, the bacteria are grown in the water and intimately mixed with the organic contaminants in the water. In a fixed-film system, bacteria are grown on an inert support medium within the reactor and the water containing the contaminants passes over the film of bacteria. Submerged fixed-film designs place the medium below the water level, and activated carbon can serve a dual role of adsorption and support media. The miscellaneous designs are the many special designs developed during the last several years.

Costs of ex situ treatment of contaminated groundwater cannot be estimated easily. The nature of the contaminated groundwater will determine which type of ex situ treatment is used and the costs will be highly contaminant and vendor specific, depending on the equipment required to build the reactor.

3.5.2.1.1 Suspended-Growth Reactors The simplest bioreactor system for treatment of groundwater is an aerated lagoon or basin. An existing pond or tank can be used as the reactor. In some cases, portable swimming pools have been used as the aeration tank. Figure 3.17 shows the configuration of an aerated lagoon. The contaminated groundwater enters the aerated vessel and bacteria in the reactor degrade the contaminants and create new biomass. The liquid residence time in the reactor, which is equal to the bacterial residence time, must be sufficient for the bacteria to reproduce.

Oxygen is supplied to the tank by a surface aerator or air diffusers. Sufficient power must be supplied to provide an adequate oxygen concentration, 2 mg/L, and/or to keep the tank completely mixed. With reactors having low residence time, oxygen supply is usually the controlling factor for bioreactor performance, whereas mixing is usually the controlling factor for those having long residence time. A 2-day residence time is the minimum recommended to maintain low concentrations of effluent contaminants, and two days is about the time required for the bacteria to reproduce and replace the biomass lost in the effluent. At lower residence times, the bacteria are washed out of the reactor such that insufficient numbers of bacteria remain for efficient removal of contaminants. At longer residence times, total flow

Figure 3.17
Aerated Lagoon

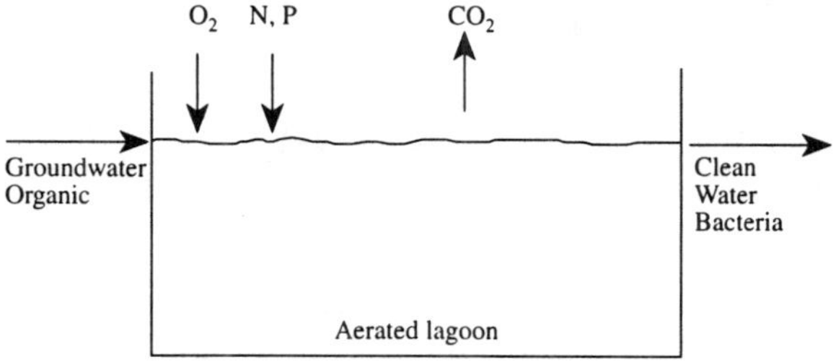

and the cost of power to mix the tank will be important operational factors. The main operating cost (other than personnel) for any aerobic biological treatment system is that of the power required for oxygen supply or mixing. These costs will limit the size of the reactor.

Two problems usually arise with the use of an aerated lagoon design. First, the extent of treatment, usually 50 to 70% of the biodegradable organic content, is controlled by the limited residence time of the bacteria. The residence time is limited because the bacteria that are created in the reaction will be in the water when it leaves the reactor. Second, the presence of bacteria in the effluent also can elicit regulatory concerns. A clarifier can be added to the system to remove the biological solids; however, bacteria grown in an aerated lagoon do not settle readily.

These problems can be solved by separating the liquid residence time from the bacterial residence time. Figure 3.18 shows that by adding a clarifier to remove the bacterial solids from the water stream and returning them to the aerated reactor, the bacterial residence time is independent of the

Figure 3.18

Bacterial Residence Time With Life Cycle Influent Concentration

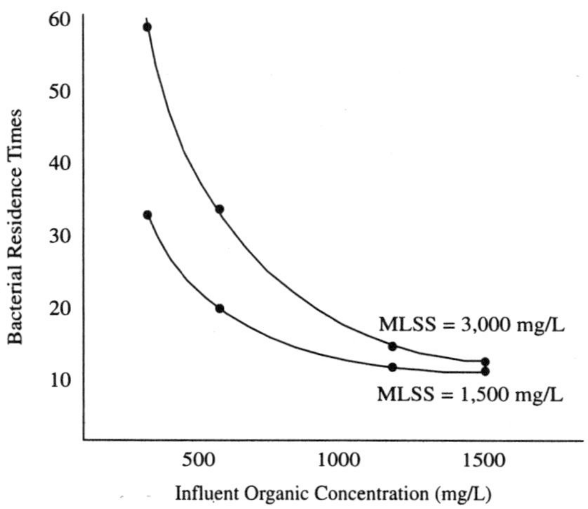

liquid retention time. From figure 3.18, the liquid residence time (R_L) is calculated as:

$$R_L = V/Q \quad [7]$$

The Bacterial Residence Time (R_B) is calculated as figure 3.19:

$$R_B = \frac{X * V}{Q_{W*} X_R + (Q - Q_W) X_E} \quad [8]$$

The bacterial residence time is governed by the wasting of the settled bacteria from the clarifier and by the uncontrolled loss of bacteria in the clarifier effluent. Bacteria returned to the aeration tank are called activated sludge (figure 3.19).

The activated sludge process is the most widely-used method of biological treatment in wastewater treatment. The basic advantages are

- the process produces low-effluent concentrations from water containing high concentrations of organic compounds;
- the system can treat many organic contaminants at the same time; and

Figure 3.19
Activated Sludge

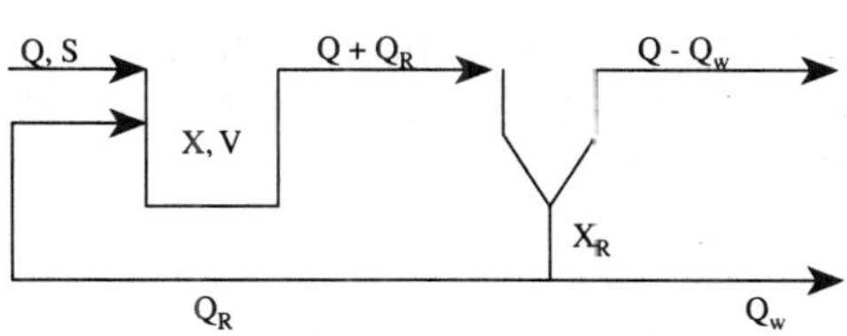

Q = Flow
Q_R = Recycle Flow
Q_w = Sludge Wastage Flow
X = Mixed Liquor Suspended Solids
X_R = Clarifier Underflow Solids Concentration
X_E = Effluent Solids Concentration
V = Volume of Aeration Basin
S = Organic Concentration

- the same equipment can be used to treat a variety of influent conditions (with equalization of the contaminant load to prevent overloading the system).

The main disadvantages are:

- the cost of manpower to keep the system adjusted to the influent conditions;
- the relative cost of oxygen transfer compared to fixed-film systems; and
- the critical need to keep the bacteria in a growth stage in which their settling tendencies are at a maximum.

The activated sludge process can remove 85 to 95% of the biodegradable organic content from the influent, and 99+% of specific organic compounds, depending on influent concentrations. Effluent concentrations of 10 to 30 mg/L of biochemical oxygen demand (BOD), a general measurement of biodegradable organic material in wastewater, can be expected with a properly operated system. Effluent concentrations of specific compounds will vary. For example, effluent phenol concentrations from an activated sludge system can be as low as 0.01 mg/L.

Certain compounds (e.g., sugars and alcohols) will degrade very quickly in a biological system. Other compounds may require longer retention times to degrade. The more readily a compound can be assimilated by the bacteria, the faster and more efficiently the bacteria can incorporate the compound into new biomass. Another approach to understanding retention time is to realize that the bacteria must first remove the easily degradable contaminants before the enzymes necessary to degrade the refractory compounds are induced. In the design of the treatment plant, this can be represented by the Food to Microorganism ratio, F/M. From figure 3.19 (on page 3.123), the formula would be:

$$F/M = Q*S/V*X \qquad [9]$$

In the activated sludge process, all of the compounds are being degraded at the same time in a completely-mixed tank. The bacteria residence time (R_B) or F/M models can be used to understand how to accomplish this concurrent degradation. One of the main advantages of the activated sludge process is that the bacterial residence time and the F/M ratio can be controlled to accommodate the degradation of a variety of compounds that have different degradation rates.

The main problem with the activated sludge design is the critical need to keep the bacteria in a form in which they readily settle. If the bacteria do not settle properly, the clarifier will not be able to remove them from the water stream. If the bacteria are not separated from the water stream and returned to the aeration basin, the whole process fails. This does not mean that the bacteria have lost their ability to degrade the contaminants. However, without being able to separate the bacterial residence time from the liquid residence time, the net result is that the activated sludge process is converted to a simple aerated lagoon with lower removal rates.

To maintain the settling properties, the environment in which the bacteria grow must be maintained relatively constant, and the bacteria must be grown to the proper sludge age that promotes flocculation. In groundwater treatment, the influent concentration of contaminants has very little variation on a day-to-day basis. There is normally no need for equalization (controlling the influent concentration of organic compounds to prevent overloading the system) as in wastewater treatment. The main problems using the activated sludge process in treating groundwater are the changing concentration of the contaminants during the project life cycle and growing the bacteria in their flocculant stage during the entire project. For illustration, assume figure 3.20 represents the influent life cycle concentration. The

Figure 3.20
Life Cycle Concentration From a Well at the Center of the Plume for an Organic Contaminant

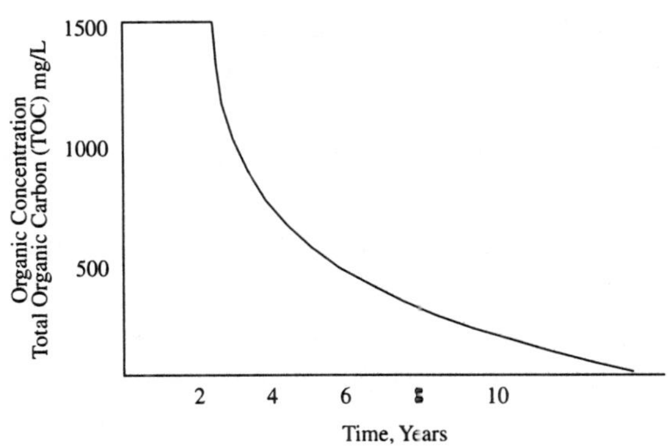

flow (Q) will be 94,625 L/day (25,000 gpd) for the entire life of the project. Also assume that all other environmental parameters are acceptable for biological treatment. Figure 3.19 (on page 3.123) will again represent the activated sludge treatment system. To keep the bacteria in a growth phase in which they settle properly, the bacteria should have a Residence Time (R_B) between 5 and 20 days. The following scenario describes what happens to the sludge age during the life of the project.

Assume:

Mixed Liquor Suspended Solids (MLSS)(X)	= 3,000 mg/L
Yield Coefficient (Y)	= 0.25 g/g (0.25 lb/lb)
Volume of the Aeration Tank (V)	= 151,400 L (40,000 gal)

$$R_B = (X*V)/(Q*S*Y)$$

For Year 1, S = 1500 mg/L

$$R_B = 12.8 \text{ days}$$

For Year 3, S = 1200 mg/L

$$R_B = 16 \text{ days}$$

For Year 5, S = 600 mg/L

$$R_B = 32 \text{ days}$$

And For Year 7, S = 300 mg/L

$$R_B = 64 \text{ days}$$

As can be seen from these data, the system will maintain the proper sludge age for about 4 years. After this time, the bacterial residence time will be too high, the bacteria will lose their settling properties, and the clarifier will not be able to separate the bacteria from the treated water. Once the clarifier fails, the system will not be able to maintain a high concentration of bacteria in the aeration basin. At this point, the system will no longer remove a high percentage of the incoming organic contaminants.

One solution to this problem is to lower the MLSS concentration. Figure 3.18 (on page 3.122) summarizes the bacterial residence times for the treatment system at MLSS levels of 3,000 mg/L and 1,500 mg/L. Lowering the MLSS concentration does extend the useful life of the treatment system, but the system still fails before the cleanup can be completed. In addition, there is a lower limit to the MLSS. The MLSS concentration entering the clarifier must be about 1,250 mg/L or above to ensure proper settling. Bacteria rely on flocculation in order to settle and a critical biomass is required to ensure enough contact between the flocculating particles.

Another method to extend the useful life of the system is to divide the aeration basin into two or more tanks. In our example, two 75,700 L (20,000 gallon) tanks, instead of the one 151,400 L (40,000 gallon) tank, could be used. Assuming 1,500 mg/L MLSS, at year 6, one aeration basin could be shut down. This would effectively halve the bacterial residence time in the system at a steady MLSS. An added advantage of this method would be that half of the blowers could also be shut down. Not only would the system last longer, but it would also cost less to run in the final years of operation.

Of course, an activated sludge system designed in this manner has limitations. The final few years of the cleanup will still create a very long sludge age. The actual design may have to include different unit operations to clean up the groundwater over the entire life cycle. Further, even with an optimal design, the activated sludge system will still require a relatively high level of operator attention to properly maintain the operation of the system.

Two other equipment designs using suspended growth bacteria are the extended aeration and contact stabilization designs (see figure 3.21 on page 3.128). Although neither of these designs has a particular advantage for groundwater treatment, both have been widely used in wastewater treatment.

The only difference between the extended aeration and activated sludge designs is that the reaction chamber (aeration basin) is larger in the extended aeration design, which extends the time that the bacteria are aerated. The performance of the extended aeration process is more stable (the larger tanks serve as internal equalization) than that of activated sludge and as a result, produces less waste sludge.

Most "packaged plants" that can be purchased from vendors are extended aeration designs. Therefore, the reader has a good chance of encountering the proposed application of this design for treatment of groundwater. When used in treating groundwater, the extended aeration design will have the same limitations as the activated sludge design.

Figure 3.21

Activated Sludge, Extended Aeration, Contact Stabilization

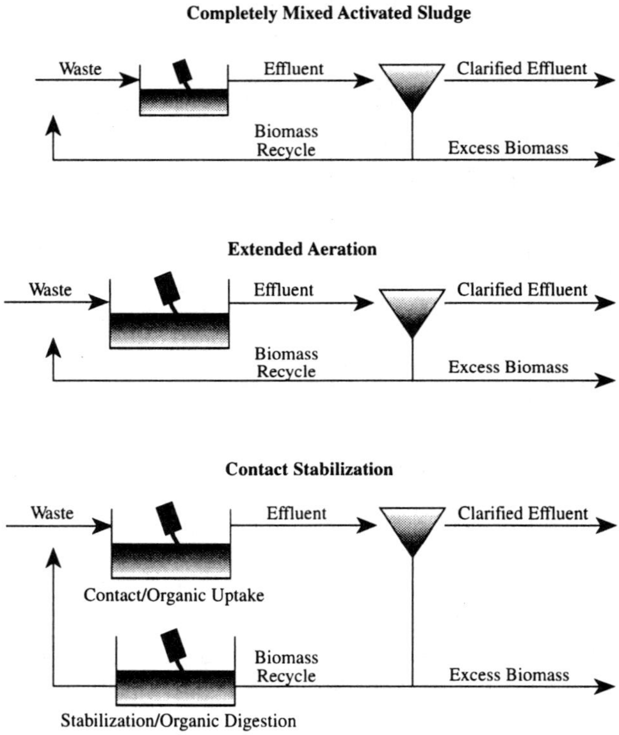

The contact stabilization process is widely used for high concentrations of easily-degradable organic compounds. The waste comes into contact with the bacteria in a small aeration tank (figure 3.21). The bacteria quickly assimilate the organic contaminants without digesting them and, with the organic contaminants, are removed in a clarifier and sent to a large aeration tank. The bacteria digest (stabilize) the contaminants in this tank. After the contaminants are digested, the bacteria are returned to the contact tank and the cycle starts again. The main purpose of this design is to save space and subsequent operating costs. It is unlikely that the reader will encounter this design in a groundwater treatment system; however, it may be mentioned in a feasibility study and knowledge of its basic design will be necessary.

One final method for applying suspended-growth bacteria is the sequencing batch reactor, which has been applied widely to industrial waste during the last several years. It also has been used to treat landfill leachate and for pond remediations. Following are the main steps in the operation (see figure 3.22):

Figure 3.22
Sequencing Batch Reactor Treatment Stages

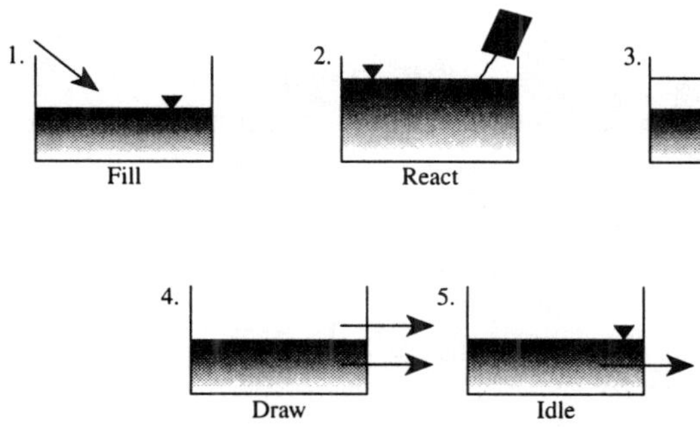

1. Fill the reaction tank with the contaminated water while maintaining full aeration;
2. Once the tank is full, bacteria completely digest the organic contaminants;
3. Stop aeration and subsequent mixing, and allow the bacteria to settle;
4. Decant the clean water and discharge; and
5. Start the cycle again.

The sequencing batch reactor design often uses two tanks operated in parallel. While one reactor is accepting water, the other reactor is going through the subsequent steps of digestion, settling, and decanting. The reactors switch back and forth to maintain a constant influent flow. The advantages of the sequencing batch reactor are simplicity of operation and flexibility under a variety of influent conditions. The main disadvantage for groundwater treatment would be operation with low concentrations of influent organic contaminants. See table 3.20 for a summary of the advantages and disadvantages of suspended-growth designs.

Table 3.20
Suspended Growth Systems

Advantages	Disadvantages
Intimate contact between biomass and waste	Completely dependent on clarifier performance
Several methods available for adjusting performance	High operation attention required
Very low concentration of specific organics in effluent	
Large-scale system relatively inexpensive	

3.5.2.1.2 Fixed-Film Reactors Another way to use bacteria in an aboveground treatment system is to set up a fixed-film biological unit. In fixed-film systems, an inert support medium with a large surface area is placed in the reactor. Bacteria naturally attach and grow to form a biofilm on any surface provided them (figure 3.23 on page 3.131). The contaminated water enters the tank and forms a thin film over the attached bacteria (fixed-film) into which the contaminants diffuse. The bacteria degrade the organic contaminants and the waste by-products (CO_2, H_2O) diffuse into the water film. Oxygen from the atmosphere diffuses through the water film and into the bacterial fixed-film. There are four important advantages in the fixed-film systems:

- the bacteria can be maintained at a high concentration without the need of a clarifier;
- oxygen can be supplied at low cost;
- the system can tolerate some variation in organic load and the intermittent presence of toxic chemicals; and
- the system is easily operated.

Figure 3.23
Fixed-Film Bacterial Growth

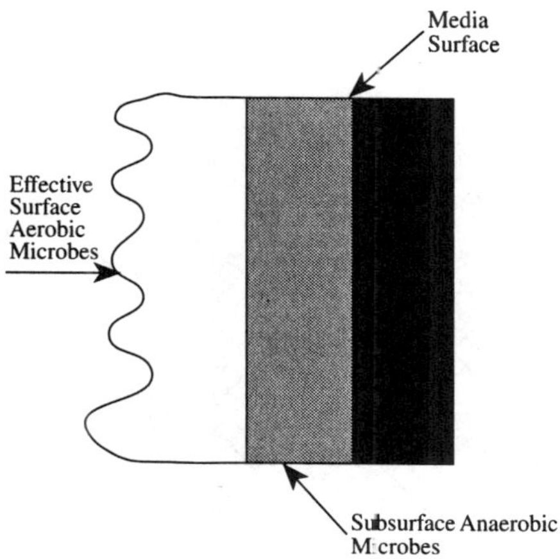

A fixed-film system requires less operator attention than an activated sludge system. Bacteria will grow attached to the medium and remove organic contaminants from the water over a wide range of operating conditions. When there are too many bacteria in the fixed-film, the bacteria will slough off and leave the reactor with the water. A clarifier can then be used to remove the biological solids before final discharge of the effluent.

The two main types of fixed-film reactors are trickling filters and rotating biological contactors (RBC). Figure 3.24 (on page 3.132) illustrates a trickling filter design. The contaminated water is pumped to the top of the reactor and distributed over the medium. The water is broken up into thin films and trickles down through the medium.

Several types of inert support media can be used in a trickling filter. Originally, small 8 to 13 cm (3 to 5 in.) diameter rocks were used to support the bacterial population. Because of the low surface area per unit volume of rock and the low oxygen transfer capacity that resulted from the small void space, only a small bacterial mass developed. Plastic media have replaced rocks in recent years. The two main categories of plastic media are dumped

Figure 3.24
Trickling Filter

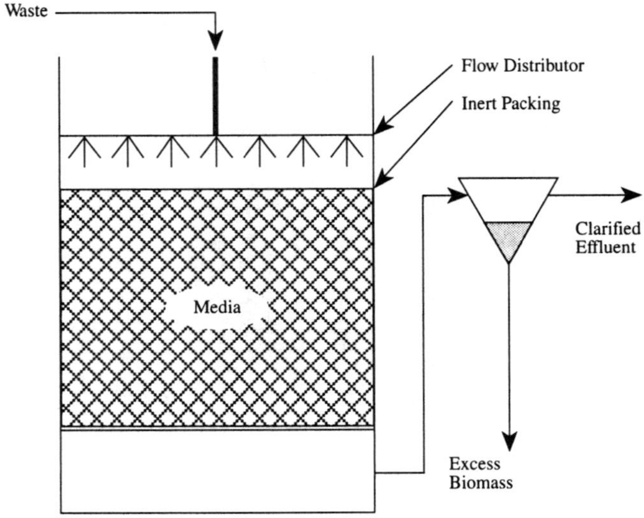

packing and stacked packing. Dumped packing is the same type of plastic medium used in packed-tower air strippers and usually is used as a replacement for rocks in existing trickling filters and small new systems. Stacked packing comes in large bricks and usually is applied to large systems.

Figure 3.25 (on page 3.133) illustrates an RBC design. The system is characterized by an elongated tank or tank series containing cylinders of plastic media, which are rotated longitudinally, as the contaminated water passes through the tank(s). In this system, the water enters one end of the tank. The medium first rotates down into the water where the contaminants come into contact with the bacteria. The medium then rotates up into the atmosphere, after which a thin film of water forms on the medium, and the oxygen transfers through the thin film of water to the bacteria. The RBC is probably the most energy-efficient oxygen transfer method for biological systems.

Figure 3.25
Rotating Biological Contactor

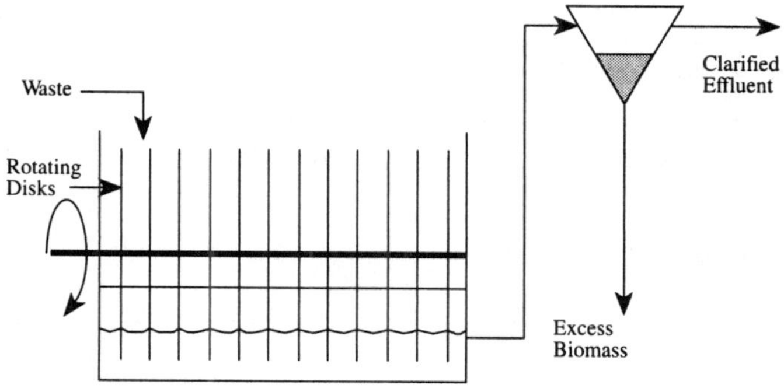

There are several technical disadvantages in the fixed-film reactors. Fixed-film reactors are plug-flow reactors. The water comes in at one end, passes by the bacterial film, and exits the other end of the reactor. As a result, the influent end of the reactor is subjected to the high concentration of the influent contaminant. In completely-mixed reactors, the influent is immediately mixed with the contents of the tank. The influent contamination may be toxic, or pockets of high concentrations of material may be found as the groundwater is recovered from the aquifer. The bacteria in the fixed-film reactor will be subjected to the full concentration. The effluent water can be recycled to minimize this effect, but this adds to the cost of operation.

Another problem with fixed-film reactors is that they will not remove as high a percentage of the influent contaminants as an activated-sludge system. Removal of particular contaminants is very important in groundwater treatment. General removal of organic compounds will be important, depending upon the final disposal of the water. The design engineer can expect 75 to 90% BOD removal and 85 to 95% removal of a specific organic compound. The lower the influent concentration, the lower the percentage removal that can be expected. Table 3.21 (on page 3.134) summarizes the advantages and disadvantages of the fixed-film systems.

Table 3.21
Fixed Film Reactors

Advantages	Disadvantages
Low operator attention	Plug flow
Retention of slow growing bacterial populations	Limited operation at high influent concentrations
Low cost oxygen transfer	Hard to adjust operation
Resistent to shock loads	
Can be shut down for up to a few weeks and then restart	

3.5.2.1.3 Submerged Fixed-Film Reactors A relatively new biological design is a combination of suspended-growth and fixed-film reactor designs, generally referred to as submerged fixed-film reactors (figure 3.26 on page 3.135). In these units, the plastic medium is submerged in the water in the reactor tank. The bacteria grow on the plastic medium as in a fixed-film system; however, the water is in constant contact with the film, as opposed to passing through in thin films.

There are two main ways in which the submerged fixed-film design can be used. In the first design, used for many years in wastewater treatment (Nyer and Ziegler 1983), the reactor is designed for completely mixed operation and for handling influents with high concentrations of organic contaminants in the influent. The medium is aerated from below, and as it is released, the air pushes water in front of it as it rises, creating an air-lift pumping action. With sufficient air, the biological reactor tank will be completely mixed. This design mode can accommodate an influent with an organic content ranging from 50 to 5,000 mg/L.

The main advantages of this design are ease of operation and high-quality performance. Submerged fixed-film reactors can perform as well as activated sludge units; however, they do not depend on a clarifier for maintaining the bacteria in the reaction tank. This design accomodates a large variety of operating conditions and requires little attention by operators. The submerged fixed-film unit combines the advantages of the suspended-growth and fixed-film systems without the disadvantages of either.

The main disadvantages of the submerged fixed-film design are the high cost of oxygen transfer and lack of scalability. Because of the nature of the design, there is a natural height limitation to the tank and, therefore, oxygen cannot be released at an optimum depth. The second problem lies in the scaling of the unit. Suspended-growth and fixed-film units become more economical as the systems get larger. Because the tank and the medium both get larger in direct relationship to the size of the system, the submerged fixed-film reactor does not provide an economy of scale for larger systems.

Neither of these disadvantages has a large effect on groundwater applications. First, the cost of oxygen transfer is a small part of the total cost of a groundwater biological treatment system. Second, most groundwater treatment systems are relatively small and the cost advantage of large-scale systems does not apply.

A second mode in which the submerged fixed-film units can be applied to groundwater is in a low-concentration design. Submerged fixed-film systems can be designed to treat influent concentrations as low as 1 to 20 mg/L, a very important consideration for groundwater applications. High concentrations (greater than 50 mg/L) are rarely found in groundwater, and

Figure 3.26
Submerged Fixed Film

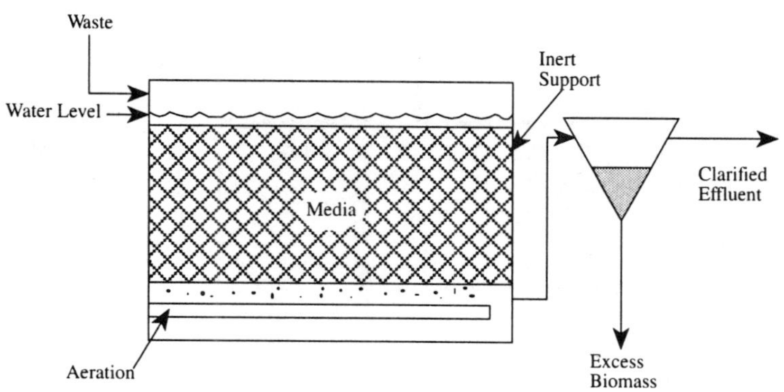

when they are, life-cycle design considerations reduce the concentration below 50 mg/L in a short period of time.

Figure 3.27 illustrates the low-concentration design of a submerged fixed-film unit. The basic design is the same as the original submerged fixed-film design. Plastic media are submerged below the water level in the reactor tank. The low-concentration reactor uses small amounts of air and a plug-flow pattern. The water enters the top of the tank and is distributed across the media. The water flows down through the media and exits the bottom through a collection system. The air is released below the media. Very small amounts of air are used due to the low oxygen demand in a low-concentration reactor and the need to maintain a nonmixed state in the reactor.

Even under these conditions, the low concentration of organic contaminants in the influent is not sufficient to maintain microbial growth in the reactor. Normally, influent concentrations of less than 20 mg/L will result in a rate of bacterial decay that is faster than bacterial growth. Therefore, the low-concentration reactor must operate in a decay mode, not in the normal growth mode of biological treatment systems. In the decay mode, bacteria usually are grown on the fixed-film using a synthetic feed source until the population size stabilizes. Then, the synthetic feed is removed and the

Figure 3.27
Low Concentration Submerged Fixed Film

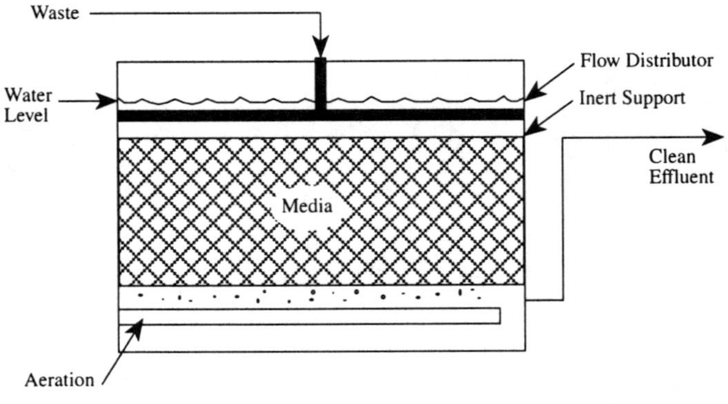

low concentration of influent is conveyed through the system. Under these conditions, the bacteria slowly decay. With proper design and operation, the decay period can last between 6 months to 1 year before regrowth is required.

There are more than 20 low-concentration reactor units currently operating full scale in the field. Compounds that have been treated in this reactor design range from acetone and methylethylketone to benzene and chlorobenzene. Reactors are also currently being used to treat groundwater containing tetrahydrofuran and *t*-butanol.

3.5.2.1.4 Reactors Based On Activated Carbon There are several other types of designs that have been used to treat organic contaminants in groundwater. Two of the most popular in recent years are the powdered activated carbon treatment (PACT) units and fluidized-bed reactors.

The PACT unit is basically an activated sludge treatment system with powdered-activated carbon maintained in the reactor (figure 3.28). The combination of powdered-activated carbon and active bacteria increases the removal capabilities of the treatment system in comparison with systems using either alone. The powdered-activated carbon can remove slowly degrading or nondegrading organic material from the water while the bacteria can attach to the powdered-activated carbon and consume the contaminants adsorbed to the carbon. The system has been used widely to treat

Figure 3.28
Powered Activated Carbon Treatment

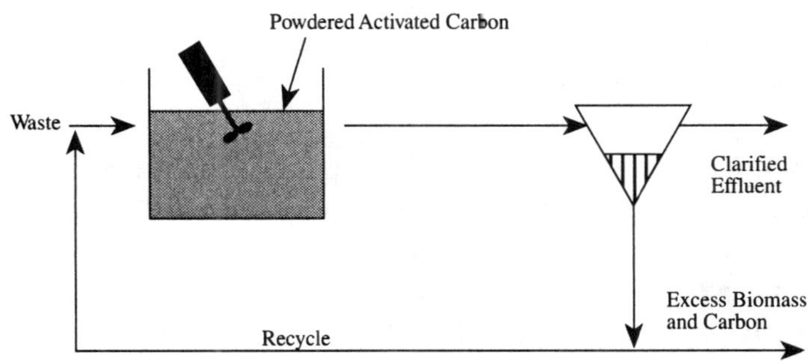

hazardous organic waste. While it has been used mainly in the wastewater field, the PACT system is now being used to treat contaminated groundwater. The main advantage of the system is that it can treat a large variety of organic compounds. The main disadvantage is that it is basically an activated-sludge design, and will suffer the same limitations.

A fluidized-bed reactor is basically a submerged fixed-film system. The fluidized-bed design has been applied recently to a number of groundwater remediation problems, including treating tank bottom wastes, produced water brines, and groundwater contaminated with BTEX, aliphatic hydrocarbons, and PAHs from gas and oil production activities. The granular activated carbon fluidized-bed reactor has been able to consistantly provide removal of 99+% for BTEX, and 2- to 4-ring PAHs for all these streams at liquid residence times of minutes and organic loading rates of 1 - 5 kg COD/m^3-d. The technology has been applied at over 40 full-scale and pilot-scale installations within the United States. Currently, the technology is being demonstrated for the removal of chlorinated solvents by operating the system in an anaerobic mode, with detention times under an hour. In this design, the support medium consists of small-diameter particles. Water and air flow in an upflow pattern through the medium, fluidizing the bed. Typical media are sand, activated carbon, glass beads, and other small particles. Recently, activated carbon has been the main medium applied in fluidized-bed reactors. The activated carbon once again fulfills the dual purpose of adsorbing organic contaminants and acting as an attachment site for bacteria.

The fluidized-bed design can also be run in high- and low-concentration modes. If the influent organic concentration is high (50 to 5,000 mg/L), high ratios of recirculation are used with pure oxygen as the oxygen source. When influent organic concentrations are low, recirculation is kept to a minimum and the liquid residence time in the reactor is kept short (5 to 30 min). The fluidized-bed design has recently been applied to several full-scale groundwater cleanups.

3.5.2.1.5 Miscellaneous Reactors Another possible approach to treatment of contaminated groundwater is anaerobic reactors. All the current laboratory work in this area indicates that bacteria can be used to degrade chlorinated hydrocarbons. Once these reactions are well understood, and the required environment defined, full-scale reactors will be possible. Several anaerobic reactor designs are currently available, mainly used to treat wastewater with high concentrations of organic constituents. The main areas of application are the food and beverage industries. Full-scale data on the application of anaerobic processes to groundwater are limited. But, as the laboratory research leads to pilot-scale applications and the technology is advanced, anaerobic reactors should be considered as an alternative treatment.

Several other types of biological reactors are also available, and new designs are constantly being developed. But with the exception of the low concentration submerged fixed-film design and the activated carbon fluidized-bed design, there are no other units that are currently being specifically designed for treatment of groundwater. Again, it is very important to understand that just because activated sludge is not a viable design for treatment of groundwater, it does not mean that biological treatment is not possible. Any new design for groundwater treatment will have to address the specific problems of groundwater. The main problem areas are life-cycle design, treatment of low concentration of organic contaminants, and portability. The designer will have to understand the advantages and disadvantages of each biological reactor and exploit them to the best advantage during the life cycle of the project.

3.5.2.2 Ex Situ Treatment of Contaminated Soils and Sediments

Contaminated soils and/or sediments are frequently present at a site where the groundwater quality has deteriorated. Bioremediation of soils may be performed either in situ or ex situ. Some sites are not suited to in situ treatment because of hydrogeological conditions at the site or the characteristics of the wastes. At such sites, soils may be remediated using ex situ treatment. This section addresses the various types of bioreactors used in ex situ treatment.

Ex situ biotreatment reactors for soil remediation fall into two main categories, slurry-phase treatment and solid-phase treatment. In slurry-phase treatment, contaminated soils or sludges are maintained as an aqueous slurry. Solid-phase biotreatment relies on principles applied in agriculture and in the biocycling of natural compounds: its process options include land treatment, soil-pile treatment, and composting.

There are advantages and disadvantages to each of these designs, but all of the reactors follow the basic concepts of any biological reactor. Their main purpose is to maintain an environment in which the organic contaminants and bacteria can react under optimum conditions. The bacterial requirements of environment (O_2, pH, temperature, etc.) and nutrient addition (NH_3, PO_4, etc.) must be met by all reactors. The solid-phase reactors have one more requirement. They must also maintain the proper moisture content.

3.5.2.2.1 Slurry-Phase Biotreatment Slurry-phase biotreatment may be performed either in bioreactor vessels or in lined lagoons, but the basic units include aeration and mechanical mixing equipment and, sometimes, an emissions-control system. Depending on the setting, slurry-phase bioremediation may be compared with the activated-sludge process or aerated lagoon treatment. In either application, soil or sludge and nutrient-amended water are combined to form an aqueous slurry. Mixing must be sufficient to keep the solids in suspension and oxygen must be supplied throughout the slurry matrix to promote aerobic microbial activity. Bioslurry reactors are operated so as to maximize mass-transfer rates and contact time between the contaminants and microorganisms.

Oversized material must be removed. The first step in the treatment process is to slurry the soil or sludge to be treated, which is passed through a trommel screen to remove gravel and debris with diameters larger than 0.64 cm (0.25 in.). More water may then be added to obtain the desired slurry density before bioslurry treatment. Maximum treatment efficiencies are generally obtained with soil slurries containing 30 to 50% dry solids by weight, although difficulty in maintaining the solids in suspension limits the acceptable slurry solids content range to 20 to 30% for some bioslurry reactors (Stroo 1991; Brox and Hanify 1989).

Three pilot-scale demonstrations of slurry-phase treatment, each lasting 60 days, were performed at a site contaminated with oil-refining wastes. The contaminants of concern were PAHs and oil and grease (Stroo 1989). The treatment vessels ranged from a 64,345 L (17,000 gal) reactor to a 2,800,000 L (750,000 gal) aerated lagoon. The reactor solids loadings ranged from 5% to 30% (i.e., the slurries consisted of 5 to 30% dry solids by weight). Data pertaining to the oil and grease treatment results were not reported, however, reductions in total PAH concentrations ranging from 76 to 92% and in carcinogenic PAHs (i.e., 5-ring and 6-ring PAHs) concentrations ranging from 25 to 89% were effected. The highest removal efficiencies were expected in the reactors with the lowest solids loadings, but this was not the case. In fact, the highest total and carcinogenic PAH reductions were obtained in the lagoon with 30% solids loading (Brox and Hanify 1989).

Slurry systems have an economic and technical advantage when the contaminated solids already contain a high-moisture content. Lagoon bottoms are a prime example. In addition to the physical advantages of not having to add water, the reactors actually perform two functions: normal degradation of the contaminants and reduction in volume, accomplished through direct degradation, and by breaking of the emulsion with the subsequent release of soils and water. Even when the soils require further treatment, the reduction in volume can still economically justify slurry treatment.

3.5.2.2.2 Ex Situ Land Treatment Ex situ land treatment is sometimes known as prepared-bed or on-site, land-based bioremediation treatment or land farming. The process involves spreading wastes over the surface to enhance natural microbial degradation of contaminants. Land treatment was the first method used for bioremediation of soils and sludges and has been successfully applied by the petroleum industry in the managed disposal of petroleum refinery wastes for decades (Hildebrandt and Wilson 1990).

An ex situ land treatment unit operates under the same conditions as in situ land treatment unit, with the possible addition of a leachate collection system installed beneath the unit. The collected leachate may be returned to the bed for additional treatment or treated by an on-site wastewater treatment system and then returned to the unit (leachates having a high-content of specific chemicals), or sent to an off-site wastewater treatment facility, if available. An ex situ land treatment unit provides better control of any leachate and, therefore, overcomes one of the potential disadvantages of an in situ land treatment unit.

A lined land treatment facility may be required for treatment of certain wastes, sludges, and contaminated soils and their constituents. The liner prevents migration of contaminants to underlying soils and/or groundwater. The subgrade construction of a lined facility is similar to the subgrade construction of an unlined facility, except that either a clayey soil of a specified compacted permeability or a synthetic (generally high-density polyethylene (HDPE)) material is used. The liner, which prevents downward migration of water, is placed across the subgrade and the sides of the berms. A leachate collection system is used in combination with the liner system to control surface drainage.

Leachate collection systems include a drainage bed, usually made of sand, generally 30 cm (12 in.) thick and sloped to a sump for discharge (figure 3.29). The sump is filled with cobbles (usually of 15.2 cm (6 in.) diameter), with HDPE or PVC discharge pipes to either a retention pond or an approved discharge facility; however, the leachate could be recycled back to the land treatment unit, particularly if water is needed for moisture control. Generally, at least 30 cm of sand or a permeable soil should be placed on top of the leachate collection system. This sand or soil acts to protect the leachate collection system. About 30 to 46 cm (12 to 18 in.) of native topsoil is then placed on the sand and acts as the initial zone of incorporation. Once the topsoil is in place, the contaminated soil is applied and incorporated in the same manner as that of an unlined facility.

Figure 3.29
Ex Situ Land Treatment

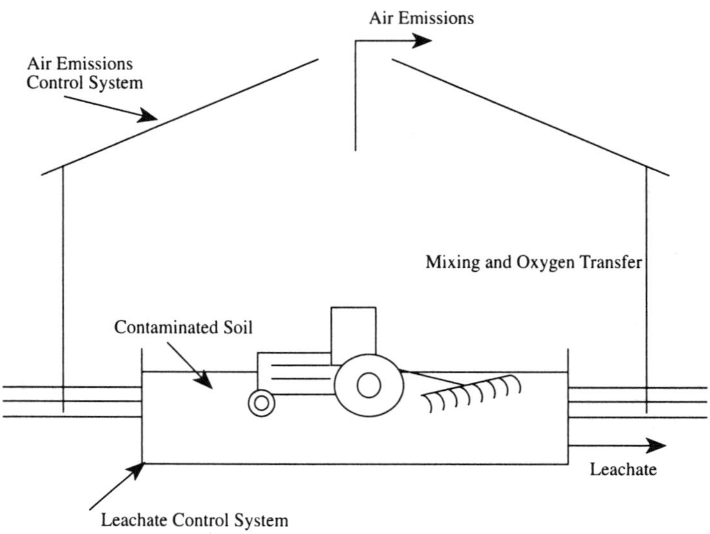

At ex situ land treatment units, the sludge or contaminated soil is commonly applied to the existing soil or prior soil-waste mixtures in lifts. A rule of thumb is to apply wastes, sludges, or contaminated soils on the ex situ unit only as deep as available tilling equipment can incorporate it. Another application, or lift, is applied only after the previously-applied material has been bioremediated to desired clean-up requirements. The unit includes a spray irrigation system for application of nutrients and inoculum, if desired, and for control of the soil moisture content. These items can also be added by the same soil mixing equipment that is used for oxygen transfer. A greenhouse top and an air-management system may also be included if containment of volatile emissions is required. The contaminated soil is tilled regularly (daily, every other day, or weekly depending upon oxygen requirements) to promote homogenization of the soil and increase the oxygen available to the indigenous mlcroorganisms. An ex situ unit can be used for many years.

The ex situ treatment site should be selected so as to minimize earth moving and grading and the volume of potentially-contaminated runoff and reduce treatment time to minimize the number and depth of lifts to be treated. Other important considerations include proximity to irrigation water, power sources, and an approved water discharge location, if needed.

The treatment unit usually is constructed adjacent to the site requiring bioremediation to minimize transportation costs and to provide better technical and management control of the process. Except for the fact that the on-site unit is constructed aboveground, the fundamentals and operation of the ex situ unit are the same as those of the in situ soil unit.

Ex situ land treatment units have been used successfully to remediate (1) contaminated soils at spill sites, (2) industrial wastes and residues, and (3) soils at surface impoundments and lagoons that are being closed. It is commonly used today to treat soil contaminated with petroleum and wood-preserving wastes. At a pilot-scale land treatment facility constructed at an oil gasification site, successful bioremediation of approximately 4,600 m^3 (6,000 yd^3) of soil contaminated with coal tar was achieved. The contaminated soil was placed in a bed to a depth of 0.6 m (2 ft) and regularly tilled and irrigated. Results achieved in a 4-month treatment period included a 73% reduction in BTX concentrations, a 36% reduction in the concentration of oil and grease, and an 86% reduction in the concentration of total PAHs. Two-ring and 3-ring PAH concentrations decreased by 92%; 4-ring PAH concentrations decreased by 80%; and 5-ring PAH concentrations decreased by 65% (Hutzler et al. 1989).

Bioremediation of soil contaminated with wood-preserving wastes was successfully demonstrated in another pilot study. The contaminants monitored were pentachlorophenol (PCP) and creosote, which is a complex mixture containing a number of PAHs. During a 5-month period, the concentration of PCP was reduced by 95%, while reductions in PAH concentrations ranging from 50 to 75% were achieved (Linkenheil and Patnode 1987).

3.5.2.2.3 Soil-Pile Treatment There are two types of soil-pile reactors. One delivers oxygen and nutrients by water movement through the soil (water-based). In the other, the air-reactor design, the nutrients are mixed in with the soil when the pile is created, and oxygen is delivered by air movement through the soil.

The water-based soil-pile treatment is identical to ex situ land treatment, except that the soil is not tilled. The contaminated soil is spread on a lined treatment bed equipped with a drainage collection system. An irrigation system is used to deliver a constant flow of a solution containing nutrients and an inoculum, if desired. The collected irrigation stream drains to a sump, from which it may be conveyed to a liquid-phase bioreactor. The bioreactor effluent is channeled back through the irrigation system. The soil-pile system may be totally enclosed if volatile emissions control is necessary.

Like ex situ land treatment, the first soil-pile treatment was successfully applied to soils contaminated with petroleum and wood-preserving wastes. At one wood-preserving site, a water-based soil-pile reactor was constructed within an existing Resource Conservation and Recovery Act (RCRA) impoundment area. Approximately 918 m^3 (1200 yd^3) of sludge and contaminated soils were mixed with an equal volume of native soils and spread in a 15 cm (6-in.) layer in the reactor. The soil-pile was irrigated daily to maintain the desired moisture content within the bed.

The contaminants monitored were benzene-extractable (BE) hydrocarbons and 16 PAHs. Concentrations of BE hydrocarbons were reduced by 69% over the first year of operation, with most of the reduction occurring during the first 4 months (i.e., May through September). Average removal rates of 95% were achieved in the first year for the 2-ring and 3-ring PAHs.

The average total PAH reduction rate was 90%, while the rate for 4- and 5-ring PAHs was 72%. As for the BE hydrocarbons, the greatest PAH reductions were achieved in the warm-weather months. These results were expected since the rate of biological degradation would be greatest during warmer weather. Further contaminant reductions achieved over the winter months were slight (Williams and Ziegenfuss 1989; Bourquin 1989; Golueke and Diaz 1989).

The air reactor design is more versatile than the water-based design, which is limited in size and oxygen transfer because of the reliance on water movement. The air reactor can be larger and handle higher concentrations.

The air soil-pile is constructed in a similar fashion to the water design (figure 3.30). An impermeable layer is placed on the ground. Then the soil (mixed with nutrients and inoculum, if required) is placed directly on the liner. There is no need for a leachate collection system since water is not added. Air pipes, placed in the soil as the pile is created, are used to deliver and/or collect air. The spacing of the pipes is dependent upon the permeability of the soil. The pipes are connected to the vacuum side of a blower. The exhaust air can be treated if required. The pile can be any size, but is

Figure 3.30
Soil Pile Reactor

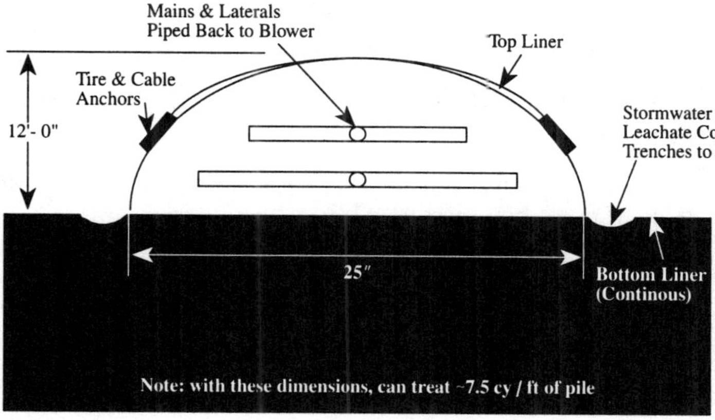

usually limited to a maximum cross-section of 3.7 x.6 m (12 x 20 ft) height and width. A plastic liner also is placed on top of the pile to keep rainwater out and to aid the proper air movement through the pile.

3.5.2.2.4 Composting *Introduction.* Composting is similar to the process used for composting of leaves, garbage, and food-processing residues. Biodegradation of the organic contaminants occurs within the compost matrix, which consists of the contaminated material mixed with organic carbon sources and bulking agents, such as straw, bark, or wood chips. At the completion of a bioremediation composting process, the treated material must be disposed of in an environmentally-sound manner. Both degradation and immobilization, however, have occurred in this process, and the composted material should not cause surface or groundwater problems at its ultimate disposal site.

The essential elements of composting are the same as for any bioremediation process:

- moisture;
- aeration;
- acclimated organisms;
- satisfactory carbon-nitrogen-phosphorus balance; and
- nontoxic conditions.

Following are the characteristics of contaminated soils or residues considered suitable for composting:

- constituents able to be eliminated by volatilization or degradation, or immobilized in the system;
- a low amount of free liquid so that an aerobic condition can be maintained;
- a high ratio of inert solids to biodegradable organic compounds; and
- a mixture that can be easily broken up by mechanical turning and/or is porous so as to allow air to move through the composting solids.

Typical composting systems that can be used for bioremediation are the windrow, in-vessel, and Beltsville systems (table 3.22 on page 3.147). The

Table 3.22
Characteristics of Different Types of Composting Systems

Type	Open/Closed	Aeration	Examples of Wastes Treated
Windrow	Open	Turning pile	Wastewater sludge, food processing wastes, manures
Beltsville	Open	Distribution system	Sludge, wastewater sludge, other organic wastes
In-vessel	Open/Closed	Distribution system	Wastewater sludge, food processing waste

windrow system is an open system in which the compost is stacked in elongated piles (windrows) and mechanically turned for aeration. In-vessel composting entails placing the compost inside an open or closed reactor in which aeration is achieved by mechanical mixing and/or blowers. The Beltsville system is an open pile with an air distribution system under the pile. Air is drawn through the pile from the atmosphere and exhausted through a blower, generally, to an air pollution control system. Bulking agents are commonly added to increase the porosity and assist the flow of air to maintain aerobic conditions. For bioremediation composting systems, adequate operational control, as well as control of all emissions such as leachate and offgases, is important.

Operational Considerations. Composting is typically a batch biological process used to treat material with high concentrations of biodegradable organic compounds. Waste destruction and conversion are achieved by thermophilic aerobic microorganisms that occur naturally in decaying organic matter. The basic objective of this process is to maximize microbial activity, while decreasing waste volume, odors, and aqueous side streams.

The elements of effective composting are adequate nutrients, satisfactory pH and temperature, nontoxic conditions, adequate oxygen, and satisfactory moisture conditions.

For some hydrocarbon-contaminated soils, the addition of municipal sludge or agricultural materials may be required to increase the temperature of the pile in order to increase the metabolic rate of the microorganisms and, thereby, increase rates of contaminant biodegradation.

Composting is a four-step process operated so as to achieve a high degree of treatment or remediation. The first step entails mixing waste or contaminated soil with a bulking agent to enhance aerobic conditions. Increasing porosity enhances oxygen transfer and decreases the moisture content of the mass. In the second step, air, and possibly heat, is introduced to the system to effect the temperature conditions necessary for aerobic decomposition of the organic compounds. The third step is curing in which biodegradation is effected. The fourth step entails separation of the bulking agent from the remediated material for reuse.

All composting processes require staging and treatment areas. These areas are typically lined with concrete to permit ease in handling materials. Area requirements are specific to the particular process, with the windrow system requiring the greatest area for soil treatment.

Temperatures within the compost should range between ambient and 50°C (120°F). This range should be maintained to maximize bioremediation, but the higher temperatures may not be possible with wastes or contaminated soils containing low concentrations of biodegradable organic compounds.

The windrow entails the deposition of wastes in 0.9 to 4.3 m (3 to 14 ft) wide windrows to permit biological decomposition. Microbial activity is facilitated by periodic turning of the windrows to aerate the mass and release excess heat, which can be detrimental to microbial growth. Fans may be used to create an induced draft through the pile. Bioremediation may be complete in approximately 6 to 8 weeks, depending on the type of organic compounds to be remediated.

All windrows should consist of at least 40% solids to maximize windrow stability and assure proper air space for aeration. Modifications to the mass can be made by adding bulking agents. A leachate runoff collection and treatment unit should be part of the windrow system, since relatively small amounts of water are removed by other mechanisms.

In static-pile composting, air is drawn through a static pile when a vacuum is generated by perforated pipes beneath the waste. Drawing air through the pile enhances aeration and reduces odor and volatile organic emissions.

A modification to this process involves static-pile aeration with pressure ventilation and controlled heat removal. Heat output, temperature, ventilation, and water removal can be monitored for process optimization. An air compressor draws a vacuum at the bottom of the pile and returns the air at the top, providing aeration. Makeup air is added, as necessary, as oxygen is depleted. Nutrients and water can be provided through an irrigation system located at the top of the pile. Air addition should be adequate to maintain internal oxygen levels in the 5 to 15% range.

Conventional windrow composting requires a hard, dry surface (usually concrete) and large tractors to turn the piles. Static and aerated piles include perforated pipes installed beneath the piles with blowers used to create a vacuum and offgas treatment. Conventional windrows and static piles may have a roof placed over the piles to reduce precipitation infiltration.

Materials that are amenable to the composting bioremediation process include sewage sludge, soils contaminated with diesel fuel and similar petroleum products, and wastes and residues from brewing, antibiotic, fermentation, food processing, mineral oil, and munitions operations. Bulking agents may be added to increase the porosity and facilitate aeration. Materials used as bulking agents include fibrous plant material, wood chips, and bark. For compounds that are difficult to biodegrade, such as those found in munitions residue, the waste can be mixed with highly biodegradable material. This material serves as a carbon source for the microorganisms while the compounds are biodegraded through cometabolism. Materials used for this purpose should be inexpensive; examples include food processing wastes, manure, and plant material.

Two field-scale studies were performed to investigate the feasibility of using composting to treat explosives and propellant-contaminated sediments at two Army ammunition plants. Contaminants of concern at one site included 2,4,6-trinitrotoluene (TNT), hexahydro-1,3,5-trinitro-1,3,5-triazine (RDX), and octahydro-1,3,5,7-tetranitro-1,3,5,7-tetraazocine (HMX). The contaminant of concern at the other site was nitrocellulose (NC) (Williams and Ziegenfuss 1989). Four compost piles of approximately 9 m^3 (12 yd^3) each were constructed at each site. The piles were placed on 8-inch thick concrete pads which drained to a sump. The compost mixture consisted of cow or horse manure, straw, alfalfa, horse feed, and contaminated sediment and was moistened with water when it was placed on the concrete pads.

Each compost pile contained a network of perforated pipes through which air was drawn to aerate the piles.

The compost period lasted 22 weeks at the first site. During that time the following results were achieved: reductions in TNT concentrations ranging from 99.6 to 99.9%; reductions in RDX concentrations ranging from 94.8 to 99.1%; and reductions in HMX concentrations ranging from 86.9 to 96.5%. At the second site, reductions in NC concentrations ranging from 91 to 98% were achieved during a 14-week composting period (Williams and Ziegenfuss 1989). Although composting was effective in reducing the concentrations of these compounds, by-products more toxic than the parent materials may be produced.

3.5.2.2.5 Cost Comparison of Ex Situ Bioremediation Technologies for Contaminated Soils Estimated per-ton costs for the four ex situ bioremediation technologies are presented in table 3.23. All unit costs are based on the assumption that the waste treated is contaminated soil having a unit weight of 1,762 kg/m^3 (110 lbs/ft^3). Ex situ land treatment and composting are attractive because of their low costs, which are in the ranges of $39 to $88/tonne ($35 to 80/ton) and $44 to $ 110/tonne ($40 to 100/ton), respectively (Lynch and Genes 1988; Environmental Protection Agency 1988; Williams and Ziegenfuss 1989). Soil-pile treatment costs are in the $99 to $110/tonne ($90 to $100/ton) range, while slurry-phase treatment costs range from $88 to $165/tonne ($80 to $150/ton) (Bourquin

Table 3.23

Treatment Reactor Costs for Solid Phase Biological Methods

Process	Cost
Off-site Disposal in Permitted Hazardous Waste Landfill	$200 to $300/ton plus transportation costs
Off-site Incineration in Permitted Facility	$300 to $1200/ton plus transportation costs
Engineered Land-Farm Treatment	$35 to $100/ton
Soil Pile Treatment	$50 to $100/ton
Composting	$50 to $70/ton
Bioslurry Reactor Treatment	$80 to $150/ton

1989; Golueke and Diaz 1989). All of the ex situ bioremediation technology costs compared favorably with the higher costs of off-site landfill disposal ($220 to $330/tonne ($200 to $300/ton)) and off-site incineration ($330 to $2,200/tonne ($300 to $2,000/ton)). The landfill disposal and incineration costs shown do not include transportation costs, which are substantial. For example, transportation costs for a 9,072 kg (20,000 lb) truckload of contaminated soil hauled by a qualified hazardous materials transporter range from $2.50 to $3.50/loaded mile.

One of the major costs of all solid-phase reactor operations is soil movement. The best way to save money on a remediation project is to incorporate the biological reactor into a project that is already moving the soil. For example, when buried drums are removed from soil, the biological system can be placed as each section is handled. The biosystem can then treat the contaminated soils at essentially no additional cost for soil movement.

3.5.2.3 Biological Reactors for Contaminated Air

3.5.2.3.1 Introduction. Biofiltration is the biological removal of organic contaminants in gas streams in a solid-phase reactor (figure 3.31). This process is well established in Europe and Japan where it has been used as an air pollution control technology in successfully controlling odors, volatile

Figure 3.31
Ex Situ Bioremediation of Volatile Organic Compounds

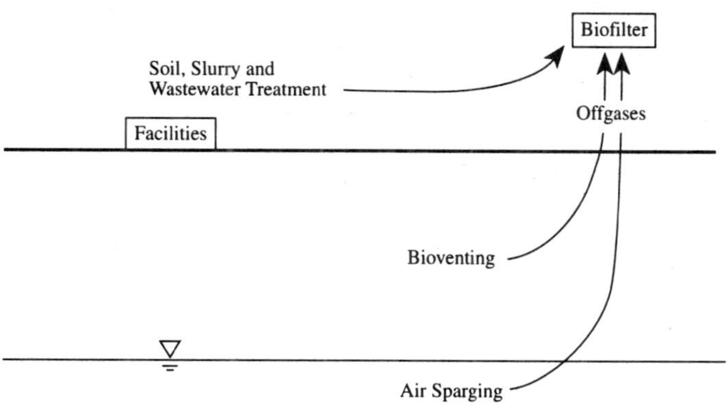

organic compounds (VOCs), and air-toxic compounds. Biofiltration has economic and other advantages over existing air pollution control methods, particularly if applied to offgas streams that contain only low concentrations (typically less than 1,000 ppm as methane) of air pollutants that are easily biodegraded (Leson and Winer 1991).

A biofilter also has been known as a "bodenfilter". A biofilter consists of one or more beds, typically 1 m (3.3 ft) in height, of biologically-active material, primarily mixtures based on compost, peat, or soil. Contaminated offgas is vented through the filter. The air contaminants diffuse into the wet, biologically-active biofilm that surrounds the filter particles. Aerobic degradation of the pollutants occurs in the biofilm if microorganisms are present that can metabolize them. End products from the complete biodegradation of air contaminants are CO_2, water, and microbial biomass. The oxidation of reduced sulfur compounds and chlorinated organic compounds also generates inorganic acids.

Compost produced from municipal waste, wood chips, bark, or leaves has been the basis of filter material used in recent applications in Europe, although peat and heather mixtures have also been used. Biofilters built in the U. S. have been mostly "soil beds" in which biologically-active mineral soils were used as filter materials.

Most biofilters have been constructed as open, single-bed systems, although enclosed, multi-story units have been considered.

In the U. S., the first systematic research on the biofiltration of H_2S was conducted in the early 1960s (Carlson and Leiser 1966). This work included the successful installation of several soil filters at a wastewater treatment plant near Seattle and demonstrated that biodegradation, rather than sorption, accounted for the odor removal. Other successful soil bed applications in the U. S. include the control of odors from rendering plants (Prokop and Bohn 1985) and the destruction of propane and butane released from an aerosol can filling operation (Kampbell et al. 1987).

Since the early 1980s, biofiltration has increasingly been used in Germany to control VOCs and air toxics emitted from industrial facilities such as chemical plants, foundries, print shops, and coating operations. This development was brought about primarily by new federal regulations (Leson and Winer 1991).

Biofilters have been shown to be suitable for many uses. They have been employed mostly to remove odors from waste gases in chemical and pharmaceutical manufacturing and food processing, as well as in storage tank vents and fuel- and solvent-handling systems (Bohn and Bohn 1988). These biofilters can remove and safely dispose of approximately 99% of slightly volatile and easily-biodegradable organic compounds, such as aldehydes, SO_2, NO_x, and H_2S, and about 90% of volatile gases such as methane, propane, and carbon monoxide. In addition, they remove liquids and solid particulates from the gas streams.

Biofilters differ from other air pollution control methods in several ways:

- contaminants are both adsorbed and oxidized;
- suitable reaction times allow complete treatment of the organic compounds; and
- both organic and inorganic gaseous pollutants are removed.

The factors limiting biofilter treatment are the biodegradability of the organic contaminants and the permeability and chemistry of media. Because these factors vary widely, the design of biofilters is specific to the particular site.

Because microbial degradation of odorous compounds and VOCs is the primary removal mechanism within a biofilter, the filter material also must provide the proper environment for microbial growth. Microorganisms require both aerobic conditions and adequate moisture. The filter should also contain materials on which the microbes can feed to ensure that the microbial population can survive if a shutdown of the entire system should occur for any length of time.

Biofilters are simple to operate and require little maintenance. The only maintenance required is merely periodic monitoring of the moisture content and pH of the filter material. The microorganisms most effective for odor removal are sensitive to acid conditions. The pH of the filter should be measured frequently at several locations and depths. If the pH drops below the recommended level, a base, such as lime, should be added to restore neutral pH conditions.

3.5.2.3.2 Operational Considerations. Humidity control in the biofilter is important. Too little moisture results in dry zones and loss of microbial activity. Too much moisture results in the development of anaerobic zones and, consequently, poor effluent quality and odor production. A moisture content between 40 to 60% by weight is considered optimal (Leson and Winer 1991). The type of distribution network is determined by two criteria: the need to distribute the gas uniformly and the power required. Three types of systems can be used:

- perforated pipe;
- pressure chamber systems; and
- cinder block systems.

In perforated pipe systems, the bed is underlaid with a network of perforated pipes set in a gravel bed. In pressure chamber systems, a large pressure chamber at the bottom of the bed supplies and distributes air. In the cinder block system, air distribution is accomplished through prefabricated, slotted, concrete blocks. The blocks provide both aeration and drainage systems.

The following are ideal properties of biofilter media:

- high adsorption capacity;
- low pressure drop;
- high nutrient content;
- pH buffering capacity;
- adequate moisture content; and
- temperature between 25°to 35°C (77° to 95°F).

The filter media provide sites for adsorption of pollutants and for attachment of microorganisms. The media also are a source of additional nutrients so that microbial activity is not limited by nutrient availability.

Other operating parameters of concern are residence time and loading rates. The removal efficiency is highly dependent on the residence time, which represents the average time that a gas molecule spends inside the filter bed. Recommended residence times for odor control vary between 5 and 30 seconds. Gas-loading rates for processes utilizing compost filter beds are typically 1.5 to 3 $m^3/m^2/min$ (5 to 10 $ft^3/ft^2/min$) (Leson and Winer 1991).

For any given offgas, the biofilter size required for the desired rate of removal of specific contaminants depends on the hourly contaminant load (in grams per hour) as compared to the degradation capacity of the filter material for a specific constituent — usually given in terms of grams per hour per cubic meter of biofilter material. Degradation rates for common air pollutants can vary widely and depend predominantly on the type of pollutant and the biological and physical characteristics of the filter material. Typical rates for easily-degradable VOCs, such as alcohols, ketones and many aliphatic and aromatic hydrocarbons, range between 50 and 100 g/m^3 per hour (3×10^{-3} to 6×10^{-3} lb/ft^3 per hour). Higher chlorinated organic compounds have lower rates of degradation, with the degradation rate decreasing with increasing chlorine substitution (Leson and Winer 1991).

Gas-flow rates affect the filter size required to provide the degradation capacity for a given pollutant load. Since offgases from industrial sources often contain a variety of organics, and since degradation rates will depend on the offgas concentration of the target pollutant, pilot testing on a partial offgas stream from a source is usually conducted to determine the required size and operational conditions for a full-scale system.

3.5.2.3.3 Costs. Installation costs for soil beds range from \$5-6 per m^3/h, and depend on the piping required to distribute the gases throughout the soil (Bohn and Bohn, 1988). The only operating cost is from the power (0.4W per m^3/h) required to overcome the 2-3 in. water gauge backpressure.

4

POTENTIAL APPLICATIONS

See also Appendix A for case studies of bioremediation and Subsection 3.2.1, Microbial Ecology and Physiology, which discusses the applicability of bioremediation to classes of compounds.

4.1 General Criteria

The usefulness of bioremediation will depend on the limits of contaminant biodegradability and the physical constraints to creating the necessary conditions to achieve an acceptable rate and extent of contaminant biodegradation. Table 4.1 (on page 4.2) summarizes the potential biodegradability of some common, naturally-occurring and xenobiotic contaminants.

Bioremediation can be used to treat sludges and water from impoundments, sediments, and groundwater within the saturated zone, soils within the unsaturated zone, excavated soils, recovered groundwater, and gases extracted from in situ soils or aboveground reactors. Remediation of a specific site can entail one bioremediation process, several in combination, or one or more in conjunction with other treatment and remedial processes. For instance, a number of sites have been treated using a groundwater recirculation system to treat the saturated soils and groundwater, a bioventing system to treat the unsaturated soils, and an aboveground soil reactor to treat excavated soils. Such systems could have easily incorporated a liquid phase bioreactor to treat the recovered groundwater as well.

The applicability of the in situ and ex situ bioremediation processes, as a function of site characteristics, type of contamination, and performance, is summarized in table 4.2 (on page 4.4). Some processes are marked with two abbreviations to indicate that specific site conditions will result in a range of effectiveness for that type of process. Primary processes are the

Table 4.1
Biodegradability of Common Contaminants

	Natural Compounds					Xenobiotic Compounds				
	Petroleum Hydrocarbons			Creosote	Low MW Alcohols Ketones Esters	Chlorinated Aliphatics		Chlorinated Aromatics		
Condition	BTEX	Low MW Gasoline #2 Fuel	High MW Oil PAHs			Highly Chlorinated	Less Chlorinated	Highly Chlorinated	Less Chlorinated	PCBs
Aerobic	1	1	2	1	1	4	1	2	1	3
Nitrate Reducing	2	3	4	2	5	5	5	5	3	5
Anaerobic	3	3	4	3	3	3	3	3	4	3

1 Readily biodegradable as growth substrate
2 Biodegradable under a narrow range of conditions
3 Transformed or partially metabolized as a second substrate (cometabolism)
4 Resistant
5 Insufficient information

main technologies used to treat the contamination. When necessary, secondary processes are used to treat the emissions and/or effluents generated by the primary processes.

Some or all of the soils at some sites may not be amenable to any ex situ technologies because excavation is precluded by the location of buildings or other structures, the depth of the contamination, or location of contaminated soils below the water table. For this type of site, in situ bioremediation may be useful.

4.2 In Situ Bioremediation

Parameters that adversely affect performance of all in situ processes include soil and sediment hydraulic conductivities (K) that are less than 10^{-4} cm/sec (3.28×10^{-6} ft/sec) and high organic matter content (see table 4.2 on page 4.4). In matrices with low values of K, it is more difficult to deliver oxygen and nutrients rapidly and pore spaces are plugged more readily than in those that are more transmissive. In addition, contaminants may sorb to organic matter and become less available for biodegradation and physical removal. Other parameters are more specific to the particular process. The effectiveness of a bioremediation process at a particular site will depend on other parameters that affect site operation and process performance, such as site and contaminant characteristics, the remediation goals, and regulatory requirements. In general, many bioremedial processes can be considered as treatment options when the contaminants are low in concentration and readily biodegradable, and when the affected matrices have K values of 10^{-4} cm/sec (3.28×10^{-6} ft/sec) or greater. Conversely, sites with high concentrations of relatively difficult to degrade contaminants and less transmissive sediments will have limited or no bioremediation options.

Although liquid delivery was the first in situ bioremedial approach developed for treating subsurface contamination, air sparging probably replaces this method at most sites where control of plume migration is not required. Air sparging is less expensive, distributes oxygen across the entire site faster, and presents fewer operational problems than the liquid delivery method. But, liquid delivery will still be used at sites where air sparging is not applicable, where a pump-and-treat system already exists

Table 4.2
Process Applicability Assuming Adequate Biodegradability of Chemicals in the Soil

Parameter	Primary Process							Secondary Process	
	Natural Assimilative Capacity	Tillage	Bioventing	Sparging	Liquid Delivery	Soil Reactors	Slurry Reactors	Liquid Reactors	Air Reactors
Surface Impoundments									
• Liquid	VL	VL	NA	NA	NA	NA	NA	H (1°)	H
• Sludge	VL	H	NA	NA	NA	L	H	H	M
Vadose Zone									
• Surficial (<2 ft)	M	H	L-H	L	NA	H	H	NA	NA
• > than 2 ft	M	NA	H	M-H	NA	H-M	H-M	NA	H
Saturated Zone									
• Dissolved	M	NA	NA	H	H	NA	NA	H	NA
• NAPLS (residual)	L	NA	NA	M	M-H	NA	NA	H	NA
Permeability									
• $>10^{-4}$ cm/sec	M	H	H	H	H	H	L	NA	NA
• $<10^{-4}$ cm/sec	L	M	M	L-M	L	M	M	NA	NA
Organic Matter Content									
• Low	NA	H	H	H	H	H	H	H	H
• High	NA	M	M	M	L	M	M	H	H
Requirement for Secondary Treatment	None or Pump-and-Treat	None	Off-Gas	Off-Gas, Pump-and-Treat may be required	Recovered Water	Off-Gas	Water Off-Gas	Off-Gas	NA
Cost									
• Capital	L	L	L	M	H	M	M-H	L-M	M
• O&M	M-H	M	L-M	L-M	H	L	H	L-M	L
Time	Slow	Moderate/Fast	Moderate	Moderate/Fast	Slow	Moderate	Fast	Fast	Fast
Limitations	Oxygen infiltration, regulatory	Volatility, accessability, space	Depth-to-water, permeability, surface access	Permeability and heterogeneity, depth-to-water	Permeability, iron, calcium, magnesium, oxygen demand vs. delivery	Permeability, access	Throughput, volatiles	Concentration or O&M costs	Mass, flow rate

Abbreviations: VL, very low; L, low; M, medium; H, high; NA, not applicable; NAPLS, nonaqueous phase liquids and liquid delivery is an easy addition and where control of plume migration is required. Sites at which air sparging would not be applicable include fractured rock aquifers, aquifers with shallow water tables (unless the surface is capped), and formations with narrow saturated intervals.

Unsaturated soils contaminated with biodegradable substrates generally can be treated by bioventing whether or not the contaminants are volatile. Bioventing is most effective where the depth-to-water exceeds ten feet and the surficial soils (upper 0.6 m (2 ft)) do not require treatment or are being treated by other methods such as land treatment. If the surface of the contaminated site is capped, bioventing can be used to treat shallower soils and sites with shallower water tables. When the depth-to-water is less than ten feet and the surface is not capped, well spacing will have to be quite close and thus capital expenditures great. Bioventing systems can also remove nonbiodegradable contaminants and those that are more difficult to degrade such as chlorinated solvents, through physical removal, provided that offgases are appropriately treated.

Bioventing is clearly more appropriate than ex situ methods, including land treatment, soil-pile treatment, slurry bioreactors, low-temperature thermal desorption, or incineration where excavation is not feasible. Where excavation is feasible, aboveground methods will be more useful than bioventing if the contaminants are difficult to treat. For moderately degradable compounds, particularly at sites containing large volumes of contaminated clayey soils, low-temperature thermal desorption may be a better remedial option than bioventing, providing excavation is feasible.

Saturated soils and groundwater can be treated in situ using variations of the liquid delivery method or sparging if the site is sufficiently transmissive. Hydraulic conductivities of 10^{-4} cm/sec (3.28×10^{-6} ft/sec)or greater are generally acceptable for these processes, although sites with lower K values may be treated if the contaminant load is light. Where applicable, in situ bioremediation will probably be the method of choice for treating aquifers contaminated with biodegradable compounds. The only other commonly applied technology is pump-and-treat which generally requires long times to meet remediation goals, except for sites that contain very soluble compounds. If timing is a concern in treating an aquifer contaminated with volatile contaminants, air sparging to transfer the volatile compounds to the unsaturated zone followed by capture with an air recovery system may be more efficient than the Raymond process; however, there will be additional costs associated with the offgas treatment.

In general, the effectiveness of in situ methods is more dependent than ex situ methods on contaminant biodegradability, contaminant concentration, and subsurface conditions. Although the transfer of nutrients and an electron acceptor to contaminant-degrading microorganisms can be easily achieved in ex situ processes, in situ methods require the movement of air or water through undisturbed soils to deliver the electron acceptor and/or nutrients.

4.3 Ex Situ Bioremediation

Excavated soils and sludges can be bioremediated with ex situ systems. Soils are most commonly treated with land treatment or soil piles. The method of choice depends on the available space for treatment, the characteristics of the contaminated soils, and the biodegradability of the contaminants. More intrinsically recalcitrant contaminants may be more effectively remediated using slurry reactors. For moderately degradable compounds, particularly for large volumes of clayey soils, low-temperature thermal desorption may be more effective than soil piles. In some instances, sanitary landfill tipping fees may be low enough to make disposal of nonhazardous soils the least costly alternative.

Slurry reactors, particularly when combined with soil washing can be used to treat a wider range of soils and contaminants than most other bioreactors. The addition and transfer of nutrients and an electron acceptor to contaminant-degrading microorganisms is more easily controlled in slurry reactors than in in situ processes.

Groundwater recovered from pump-and-treat systems, including aquifer bioremediation systems, and discharge water from bioreactors can be treated in liquid reactors. Nonbiodegradable constituents in the water to be treated can be removed in a polishing step using activated carbon or chemical treatment. Liquid-phase bioreactors are particularly effective for biodegradable organic contaminants that are also soluble and nonvolatile; such compounds are poorly removed by either air strippers or activated carbon.

4.4 Biological Reactors for Contaminated Air

Air bioreactors can be used to treat the offgas emissions from bioventing systems, in situ vapor recovery systems, soil reactors, and air-stripper towers. Air bioreactors are most applicable when the concentration of volatiles in the air phase is moderate-to-low. For systems designed for the rapid physical removal of volatile contaminants, air bioreactors might be used after the rate of physical removal has diminished.

5 PROCESS EVALUATION

See figure 5.1 (on page 5.2) for a schematic summary of an aerobic biochemical process. In this example, benzene (C_6H_6) is used by the bacteria as a sole source of carbon and energy for growth. To complete the reaction, the bacteria require a terminal electron acceptor (in this case oxygen) and major nutrients (nitrogen and phosphorus), as well as other minor nutrients commonly found in soil. In addition, the environment must be conducive to the growth of the benzene-degrading organisms (suitable pH and temperature). As a result of the benzene biodegradation, by-products (CO_2, H_2O) and new bacteria will be produced.

5.1 Basic Considerations

Proper bioremediation design involves creating the correct environment for the bacteria, and supplying the bacteria with the material that is limiting their rate of reaction (usually O_2 and nutrients). Figure 5.1 (on page 5.2) represents an aerobic design, but other biochemical reactions can be represented by a simple switch of the electron acceptor. The same basic requirements apply to any biological process, whether the reactor is aboveground, in situ, or consists of water or soils that are being treated.

The key to the design of an ex situ liquid-phase reactor is to contain the bacteria in the reaction zone. In solid-phase reactors, the delivery and expense of supplying oxygen and nutrients and maintaining the correct environment are the keys to design. For in situ remediation designs, the key is delivery of oxygen and sometimes nutrients to the bacteria in the zone of contamination and, to a lesser extent, the expense of the oxygen or electron acceptor that must be supplied. The design concerns for all of these processes center around the practical problems in maintaining the correct environment and delivering the needed material to the bacteria.

Figure 5.1
Biochemical Process

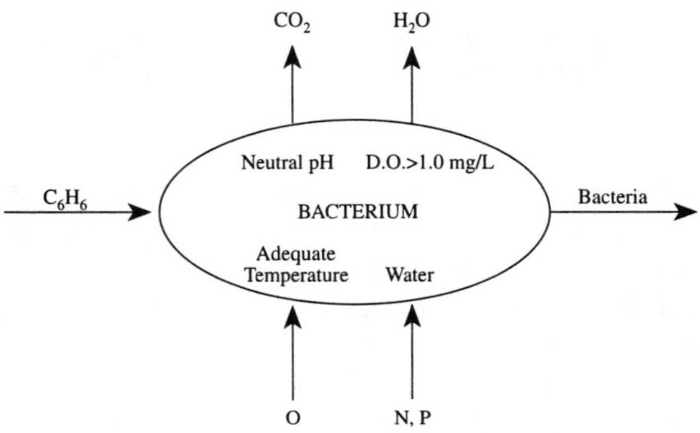

The key consideration in biological treatment is that the process is a living system that is run 24 hours a day, 7 days a week. The system cannot be turned on and off. Furthermore, biochemical processes must have a start-up period. A certain amount of time is required for the bacteria to adapt to degrade the contaminants. The last key consideration is that living systems do not respond well to severe changes in environmental conditions. Biochemical processes operate best within optimum ranges of pH and temperature.

5.2 *Process By-Products*

The by-products that result from a biochemical process depend on the original organic contaminants that are being degraded, the environment in which the bacteria are grown, and the terminal electron acceptor used by the bacteria. Most remediation designs strive to degrade the organic material to CO_2 and H_2O. New bacteria will also be produced during degradation, and the new bacteria can be considered as a by-product of the reaction. When oxygen is the terminal electron acceptor and the organic compound is degradable, then the only by-products are CO_2, H_2O, bacteria, and depending on the substrate, inorganic substituents (e.g. Cl^- and NO_2^-). But, under anaerobic conditions, other by-products can be produced. Probably the best known of these by-products comes from anaerobic degradation of chlorinated aliphatic solvents (e.g., tetrachloroethylene and trichloroethylene). Biodegradation of these compounds under anaerobic conditions in the field tends to produce lower substituted chlorinated hydrocarbons, and, as a final by-product, either chloroethane or vinyl chloride. It is important to remember that these reactions in the field are natural under certain environmental conditions. But such end products would be considered unacceptable in the design of a bioremediation process which would result in their formation.

5.3 Cleanup Levels Achievable and Duration of Treatment

Contaminant concentration often declines to a certain level during bioremediation, after which little or no biodegradation occurs. A combination of contaminant and site-specific characteristics is responsible for this asymptotic removal. Contaminant-specific characteristics that affect biodegradation include: 1) the biodegradability of the contaminant, 2) the presence of microorganisms adapted to degrade the contaminant, 3) whether the contaminant serves as a growth source for the microorganisms, and 4) the concentration of the contaminant.

Many environmental contaminants are biodegradable, at least in laboratory studies. In instances in which a biodegradable compound does not biodegrade in the field, some environmental factor is limiting or the microorganisms are not adapted to degrade the contaminant.

In cases in which the contaminant does not serve as a growth substrate, a primary substrate must be present for the microorganisms to grow on while the contaminant (secondary substrate) is fortuitously degraded (see Section 3.2.1.3, Microbial Metabolism of Contaminants that are not Growth Substrates). The rate and extent of contaminant biodegradation will be affected by the concentration of the primary substrate. For some contaminants, such as trichloroethylene, a high concentration of the primary substrate may decrease contaminant biodegradation because both primary and secondary substrates compete for the same enzyme. For other compounds, high concentrations of a primary substrate may be necessary for microorganisms to grow so that sufficient quantities of enzymes are available to effect desired reactions.

The concentration of contaminant is also important. The greater the mass of contamination, the greater the time required for cleanup. In addition, high concentrations of some compounds, such as benzene and trichloroethylene, may be toxic to microorganisms and inhibit bioremediation altogether.

Thresholds below which biodegradation of a contaminant does not occur, although it can serve as a growth source, probably exist in the field (Alexander, 1985). Essentially, the substrate may be present at a concentration that can not support growth or induce the expression of enzymes necessary to effect its degradation. Thresholds are probably organism-specific, with high nutrient requiring (eutrophic) microorganisms having higher thresholds than low nutrient requiring microorganisms (oligotrophs). In addition, the presence of other growth substrates may decrease the threshold concentration for contaminant biodegradation.

Site-specific characteristics also affect contaminant biodegradation. For both in situ and ex situ treatment, sandy materials with low organic matter content are more amenable to treatment than materials with high organic matter and/or clay content. Organic contaminants may sorb to clay and organic matter and are then rendered unavailable for biodegradation (see Section 3.2.1.2.5, Contaminant Bioavailability; Section 3.3.3 Design Considerations for In Situ Bioremediation of Aquifers). In the presence of high concentrations of organic matter or clay, biodegradation may be limited by the rate of contaminant desorption.

Contaminant biodegradation may also be affected by the permeability of the formation. Injection of electron acceptors and nutrients will be more difficult in low versus high permeability materials, thus increasing the time required for treatment and possibly decreasing the clean-up level achievable.

In instances in which the results of bioremedial efforts do not meet regulatory standards, other remedial measures may be required. For in situ processes, groundwater may be extracted and treated at the surface with a polishing step such as activated carbon or air stripping, if permitted. For in situ and ex situ processes, surfactants may be added to render sorbed contaminants more bioavailable. Although adding surfactants to ex situ processes is easily accomplished, transport of surfactants through the subsurface is more difficult. Surfactants added to the subsurface must be nontoxic and biodegradable.

5.4 Cost

The cost of bioremediation varies depending on the type and quantity of organic compounds present, site conditions, and the total volume of material to be processed.

Biological processes can be applied directly to the aquifer and vadose zone, or in reactors for aboveground systems. The main costs result from movement of the liquid or soils to the reaction zone, oxygen supply in aerobic systems, and nutrient supply. There are too many variables in all of the designs and in operational needs to give accurate ranges of cost. Each situation is unique and two designers looking at the same situation may develop completely different methods for solving the practical problems of designing and implementing the biological process.

Nevertheless, the following are generally true:

- biological treatment of groundwater or liquids from ponds is more expensive than air stripping, but less expensive than air stripping followed by air treatment or direct carbon adsorption of the contaminants. But, this generalization does not apply when very low concentrations in water (<1 mg/L organic content) are treated;
- solid-phase reactor costs relate more to the cost of materials handling and, as a second consideration, the cost of mixing supplying oxygen;
- biological treatment of soils costs less than incineration, but about the same as thermal desorption or landfilling; and
- in situ bioremediation is usually more expensive than a soil vacuum extraction system because it runs longer (usually used in

conjunction with soil vacuum extraction instead of as a substitute), and less than a pump-and-treat system.

The amount of soil moved is the key to comparing the relative costs of ex situ soil bioremediation technologies. The major cost of in situ treatment is that associated with the delivery of oxygen and nutrients to the bacteria, and it is, therefore, less expensive than an aboveground soil reactor. Because monitoring can be a major expense for in situ designs, efforts should be made to minimize the required monitoring.

6
LIMITATIONS

The key factor which controls bioremedial processes is the biodegradability of the organic waste. Provided that the waste is biodegradable, the success of bioremediation will depend on a system that is designed to maximize the rate and extent of contaminant biodegradation. Such a system is designed to deliver adequate concentrations of limiting nutrients and an electron acceptor to the contaminant-degrading microorganisms. The limitations of bioremediation are related to this delivery.

Application of bioremediation systems, like many other remediation techniques, involves some risk because the system has to operate for the full cycle time for a prognosis. Laboratory tests generally have not been accurate predictors of field remediation rates or the extent of degradation under field conditions.

6.1 General Limitations

All bioremediation processes are limited to biodegradable contaminants or mixtures of biodegradable and nonbiodegradable contaminants where bioremediation is combined with another technology. In general, soil bioremediation processes are more difficult to apply to clayey and other low-permeability soils. Sites containing biodegradable compounds might not be suitable for bioremediation if the contamination levels are high and the clean-up targets are low. This is particularly true with a contaminant or mixture that is at least moderately recalcitrant. Bioremediation can also be precluded when the contaminant is unavailable to the bacteria, such as when polycyclic aromatic hydrocarbons are strongly adsorbed to the soils or when oils are highly weathered or adsorbed by asphalt, ashes, or other material in the soil.

Limitations of each bioremediation process were indicated in the last row of table 4.2 (on page 4.4). As indicated, reliance on natural assimilative capacity is limited by oxygen and possibly, nutrient availability. These limitations are most severe when the contamination levels are high and located deep below the surface, and when the permeability is low. The practicality of natural bioremediation may also be limited by the cost of demonstrating its effectiveness to obtain approval and meet potentially stringent monitoring requirements by the controlling regulatory agency.

Land disposal restrictions (LDRs) require that RCRA wastes be pre-treated if they are to be disposed in a landfill. EPA requires that wastes be treated to a specific treatment standard based on the best demonstrated technology (BDAT), or by a specified technology. Bioremediation, along with other technologies, has been designated as a technology applicable to treatment of wastes prior to placement in a landfill. These rules are expected to increase the use of bioremediation as well as other specified technologies.

6.2 *Land Treatment Processes*

The use of tillage methods is not appropriate for treating soils contaminated with volatile constituents that create air emissions of concern or where the soils cannot be accessed or excavated. The use of tillage methods to treat excavated soils may also be rendered impractical by the lack of sufficient appropriate open space to apply the soils.

6.3 *Bioventing*

Bioventing is not applicable to surficial soils (less than 0.6 m (2 ft) from the surface) or where the water table is less than approximately 3 m (10 ft) unless the soils are capped. Clayey materials with low permeabilities usually are not amenable to bioventing; however, success has been achieved in some clayey soils. The limitations of this process in different soil types are not fully understood. Bioventing of nonvolatile, degradable compounds may be impractical at sites where nutrient addition is needed.

6.4 Air Sparging

The use of air sparging for bioremediation is relatively new and there is much to learn about its limitations. In most cases, air sparging requires vapor and groundwater recovery systems or appropriate controls and pilot testing to prevent loss of the contaminants from the treatment zone. In heterogeneous soils with clay lenses, gravel stringers, etc., channeling of the air bubbles through the permeable layers may direct the oxygen away from the contaminated zone. The area of influence of air sparging systems is related to the saturated interval or the depth below the water table at which the air is introduced. When these distances are short, the area of influence of the sparging system may be too small to be practical.

6.5 Liquid Delivery Processes

Bioremediation of aquifers using the liquid delivery process is most commonly limited by the rate of groundwater recirculation relative to the contamination level. These systems can also fail if soil pores become clogged by precipitation of iron as result of the addition of the oxygen source in high iron aquifers, by precipitation of nutrients that are incompatible with the groundwater chemistry (hard waters), or through formation of excess biomass.

6.6 Soil-Pile Treatment

Soil piles may not be applicable to clayey soils unless bulking agents are added; however, the use of slow-release oxygen compounds, such as calcium or magnesium hydrogen peroxide complexes, may overcome soil permeability limitations in some cases. This technology also requires that the contaminated soils are accessible and that sufficient land is available to treat the soils.

6.7 *Slurry Reactors*

Slurry reactors require control of off-gases when volatile constituents are present above regulated levels. This is difficult to accomplish when treating a lagoon or other large system. While biodegradation rates can be quite rapid in bioreactors, throughput is relatively small unless a lagoon or pond is used. Each type of liquid bioreactor has its own limitations as shown in table 4.2 (on page 4.4). In general, the selected system is likely to be limited by either the concentration of organic contaminants in the influent or the requirements for constant supervision.

6.8 *Air Bioreactors*

Air bioreactors, which are used to reduce emissions from primary treatment, can be limited by the size of the reactor required to treat the air flow and the mass of contaminants in the vapor phase. In some cases, the required capacity of the system can be minimized by using an alternative method, such as a catalytic converter, during the first few weeks of operation when the concentration of the recovered volatiles is highest. Alternatively, the primary system can be made fully operational over several weeks, although this practice would sacrifice the efficiency of the primary system and its ability to control off-gas treatment costs.

7

TECHNOLOGY PROGNOSIS

7.1 Application of Bioremediation

Underground storage tank releases and other sources of environmental contamination are still being found at a greater rate than sites are being remediated. Increased environmental enforcement and the large number of contractors and consultants promoting bioremediation are an indication that bioremediation will be applied at more and more sites over the next several years. Recent changes in regulations appear to remove some restrictions on excavating, bioremediating, and reusing soils on site. However, other environmental regulations may impede implementation of bioremediation processes by restricting the use of nutrients during in situ treatment and by stipulating clean-up standards that are overly conservative. Other factors affecting the use of bioremediation include competition by other remediation techniques, problems in bioremediating relatively recalcitrant chemicals, and application where difficult geological conditions prevail.

Over the next several years, commercial bioremediation will continue to focus on aerobic processes and will continue to be used widely to treat easily-biodegradable contaminants, such as low to moderate weight petroleum hydrocarbons and many oxygenated hydrocarbons, in relatively permeable soils. In situ bioremediation will continue to be attractive where the site conditions restrict excavation. For excavated soils, disposal in landfills, asphalt production, and low-temperature thermal methods will continue to compete with on-site ex situ bioremediation. However, there soon will be several soil recycling centers that use permanent soil-pile or land treatment units to remediate soils for reuse or for clean fill at municipal landfills.

7.2 Process Improvements

Improvements in the engineering aspects of bioremediation, particularly those relating to the delivery of nutrients and/or electron acceptors, will expand the conditions under which bioremediation can be cost-effective and compete favorably with alternatives. Since the late 1980s, bioventing and air sparging techniques have found increasing use as a result of improved techniques for using these methods. The U.S. Air Force has committed to using bioventing at a large number of facilities. To date, air sparging has been implemented mainly for remediation of source areas. Air sparging-induced biodegradation will also be used in place of physical barriers and pump-and-treat systems to prevent migration of biodegradable compounds and as a means to introduce cometabolites for treating chlorinated solvents. The large number of private companies and government agencies using bioremediation and conducting pilot studies indicates that there will be continued engineering improvements. The involvement of industry and agencies, such as the U.S. Environmental Protection Agency, in promoting the use of innovative technologies through testing, development, training seminars, and symposia, will promote the rapid spread of improved bioremediation processes.

Improved processes include treatment of contaminated vapors using in-place soils, as well as porous bed ex situ reactors and aqueous phase bioreactors. Natural attenuation (also known as passive or intrinsic bioremediation) will be used more widely as the ability to predict and measure rates of biodegradation improves, and regulatory agencies recognize the value of minimal disturbance to the environment and the cost-effectiveness of such attenuation.

Since the 1980s, there have been widespread efforts to introduce new and modified microbial approaches. Use of white rot fungus, cometabolic processes, anaerobic processes, and methods for improving microbial transport through porous media are being pursued by several government and private organizations. Many organizations are making rapid progress in developing new strains of genetically-engineered microorganisms that have specialized metabolic capabilities. The widespread use of such microorganisms, however, will require technical developments in transporting microorganisms and improving survival traits. The use of genetically-engineered microorganisms will also require acceptance by society and regulatory agencies.

Many of the improved processes have been shown to be effective in the laboratory and, in some cases, pilot-scale demonstrations. Their commercialization, however, will require additional effort. The small-scale results will have to be applied to real world situations and the processes will have to function routinely and continually. The question is not so much *whether* a contaminant can be treated effectively, but *when* the technology will be available. The issue of when may largely be controlled by the ability to make the necessary engineering modifications to the processes used for site-specific applications. To date, most applications of bioremedial processes have involved the use of indigenous microorganisms to treat the readily degradable constituents.

7.3 Site Characterization

Expanded use of bioremediation will depend on the ability to ask the right questions and provide appropriate answers. Improvements are needed in the methods of site investigations to provide more detailed delineation and better real-time data. Site characterizations need to provide adequate information to select and design a remedy and to follow remedial progress addressing such issues as proximity of water supplies, water table, and population density. However, generation of excessive amounts of data will be costly and dilute the evaluation and interpretation of the *essential* data.

In addition to contamination identification and concentrations, improvements are also needed in identifying microbial parameters including microbial populations and metabolic capabilities, nutrient requirements and availability, and identification of soil heterogeneity.

7.4 Site Closure Criteria

The present criteria for site closure need re-evaluation. Many sites may never reach closure because clean-up standards, particularly for total petroleum hydrocarbons, may be overly conservative and difficult, if not impossible, to achieve. Instead of imposing unnecessary conservative clean-up criteria, efforts should be focused on achieving remediation to a level that is adequate and protective of human health and the environment. Using this more realistic approach to remediation, the chemical of concern will be detoxified, degraded, and immobilized.

As these issues are resolved, bioremediation will be used to treat an ever-growing list of contaminant types under increasingly more demanding conditions. Increased information and awareness will also lead to a more scientific/technical approach to implementing and evaluating bioremediation. As a result, implemented bioremediation projects will become more reliable, especially for the less recalcitrant contaminants.

A
CASE STUDIES

BIOVENTING

There have been few well-studied examples of bioventing. The joint U.S. Air Force/ U.S. EPA RREL bioventing demonstration project being conducted at Hill Air Force Base near Ogden, Utah provides insight into the bioventing process. A catastrophic release of 100,000 L (27,000 gal) of JP-4 jet fuel contaminated the upper 20 m (60 ft) over an area approximately 4,000 m^2 (1 a) in the delta outwash of the Weber River. The depth to water was approximately 200 m (600 ft) with occasional clay stringers. Soil moisture averaged less than 6%. Jet fuel concentrations were as high as 20,000 mg/kg with an average of approximately 400 mg/kg. Measurements of total petroleum hydrocarbons (TPH) ranged to 20,000 mg/kg with an average of 1,500 mg/kg (Dupont 1992b).

The vapor recovery system consisted of vertical wells at 12 m (40 ft) intervals, which were screened from 3 to 18 m (10 to 60 ft) below the surface. Off gases were treated by catalytic incineration. Initially, the total air recovery rate was approximately 40 m^3/hr (26 acfm) or 0.04 pore volumes per day. As the hydrocarbon levels in the vent gas decreased, the rate of recovery was gradually increased to approximately 2,500 m^3/hr (1,500 acfm) or approximately 2.5 pore volumes per day. After one year of operation at the higher flow rate, the ventilation rates were reduced to between 500 and 1,000 m^3/hr (300 and 600 acfm) from wells located on the periphery of the site to maximize the retention time within the contaminated zone and to allow discontinuation of the off gas treatment while meeting the prevailing air quality regulations.

The process was monitored at several locations and depths for air pressure, oxygen, and carbon dioxide. Carbon dioxide and oxygen levels were compared to a background well that was located approximately 200 m (700 ft) from the site in the same geological unit. Oxygen consumption and carbon dioxide production were observed throughout the treatment period. The oxygen uptake rates appeared to be first-order.

Laboratory studies conducted at Battelle Memorial Institute in May, 1989, indicated that both moisture management and nutrient addition stimulated biodegradation rates in the Hill Air Force Base soils. These findings were incorporated into the field operation.

The upper 15 cm (6 in.) of soil was tilled to mix in the nutrients. Surface spray irrigation was used to provide moisture and to distribute the nutrients to depth. Soil samples at the end of the study detected both nitrogen and phosphorus throughout the range of soil depths being treated. The results of the field study showed that raising moisture levels to 30% to 50% of field capacity statistically increased respiration rates. In contrast to the laboratory findings, there was little impact from nutrient addition on rates of biodegradation in the field. However, nutrients may not have been limiting at the time of the addition because TPH levels were already below 100 ppm.

Field respirometry tests conducted during the treatment period indicated that oxygen uptake was a better indicator of biological activity than carbon dioxide production. Oxygen uptake calculations were based on stoichiometric conversion of hexane to carbon dioxide. During the high flow rate period, 15% to 25% of the decrease in hydrocarbon concentration was attributed to biodegradation. For the entire treatment period, approximately 45% of the decline in hydrocarbon concentration was due to biodegradation. However, because the hydrocarbons are assimilated into biomass, as well as metabolized to CO_2, it is likely that less than stoichiometric quantities of oxygen are required, and thus, these calculations probably underestimate the percent removal due to microbial activities.

Over the first 10 months of operation, approximately 90,000 kg (200,000 lb) of TPH was removed, reducing the soil TPH concentrations to approximately 80 mg/kg. The following 14 months of operation resulted in removal of an additional 5,500 kg (12,000 lb) of TPH, resulting in average soil TPH levels of 8 mg/kg. The overall removal efficiency exceeded 99%.

Another field bioventing study was conducted at Tyndall AFB in Florida, where JP-4 jet fuel leaked from a tank farm. This test was conducted to evaluate the effectiveness of low flow-rate bioventing as the sole treatment.

The unsaturated soils were largely sand with a depth to water of only 1.2 m (4 ft). Four plots were used in the test: two plots in the area of highest contamination where levels ranged from 3,000 to 8,000 mg/kg (dry weight) measured as TPH and two plots in a designated background location where TPH levels ranged from 95 to 140 mg/kg. Each plot was constructed by vertically installing plastic-wrapped plywood to a depth of 1.8 m (6 ft). The test plots were 2 x 6 m (6 x 18 ft) while the background plots were 1.3 x 4 m (4 x 12 ft). A series of wells were used to dewater the area around the test plots to a depth of 1.8 m (6ft). Plastic covers were placed over each plot.

Each test plot contained ground monitoring wells and three banks of multilayer soil gas sampling points. Air flows were maintained to provide 0.25 to 2.0 pore volumes of air per day. Moisture and nutrients were added at the surface via drip irrigation. The nutrients consisted of fertilizer-grade ammonium nitrate and triple superphosphate ($Ca(H_2PO_4)_2$). The application rate was at a C:N:P ratio of 100:10:1. Physical removal in the off gas was measured as hexane equivalents while biodegradation was approximated from oxygen depletion and from carbon dioxide production using a stoichiometric conversion of 3.5 kg of oxygen per kg of hexane.

Nutrient addition did not significantly enhance biodegradation. Moisture addition tests were inconclusive because climatic conditions maintained moisture levels at 70 to 80% of field capacity. Temperature fluctuations during the field test caused the soil temperature to vary from 17° to 25° C (63° to 77°F). Respiration rates increased with temperature and approximated an Arrhenius relationship (see Section on Microbial Ecology and Physiology). Variations in the air flow rate had the expected result on the relative amounts of jet fuel that were removed physically versus biodegraded. As the flow rate increased from 2 L per minute (0.5 pore volumes per day) to 8 L per minute (two pore volumes per day), the percent removed by biodegradation decreased from approximately 80 to 65%.

One of the background test plots was used as a vapor-phase bioreactor. Hydrocarbon degradation rates averaged approximately 1.93 g of TPH per day per m^3 of bed at loading rates of 2.0 g TPH per day per m^3 of bed volume. For these specific conditions, a 4:1 ratio of uncontaminated:contaminated soil was adequate for complete degradation at soil vent gas flow rates of one to two pore volumes per day. These results are important because they provided some guidelines for designing bioventing systems that could be operated with or without minimal off gas treatment.

BIOREMEDIATION OF CONTAMINATED SUBSURFACE MATERIALS USING LIQUID DELIVERY: TRAVERSE CITY, MI

(adapted from Ward et al. 1989; Fiorenza 1991; and Wilson, Armstrong, and Rifai 1993)

Introduction

In March 1988, a quantitative demonstration of the liquid delivery method for treating contaminated groundwater and subsurface materials was initiated (Ward et al., 1989). Hydrogen peroxide, which decomposes to form oxygen and water, was used as the source of oxygen for hydrocarbon metabolism. The demonstration was conducted at a site where about 3,900 L (10,000 gal) of aviation fuel had leaked from an underground storage tank at a Coast Guard Air Station in Traverse City, MI (Twenter, Cummings, and Grannemann 1985). The spill occurred in 1969 but was not detected until 1980. The aviation fuel spill as well as other sources contributed to the resulting plume (figure A.1on page A.5). Benzene, toluene, ethylbenzene, and the xylene isomers were the major groundwater contaminants. In 1985, an interdiction field was installed to intercept the plume.

Materials and Methods

Site Characterization

A field demonstration was conducted in a 9 x 30 m (30 x 100 ft) plot located upgradient and in the plume of the fuel spill. The overall hydrogeology and contaminant regime at the site had been characterized before the demonstration (Twenter, Cummings, and Grannemann 1985).

Figure A.1
Schematic Diagram of Fuel Spill and Resulting Plume

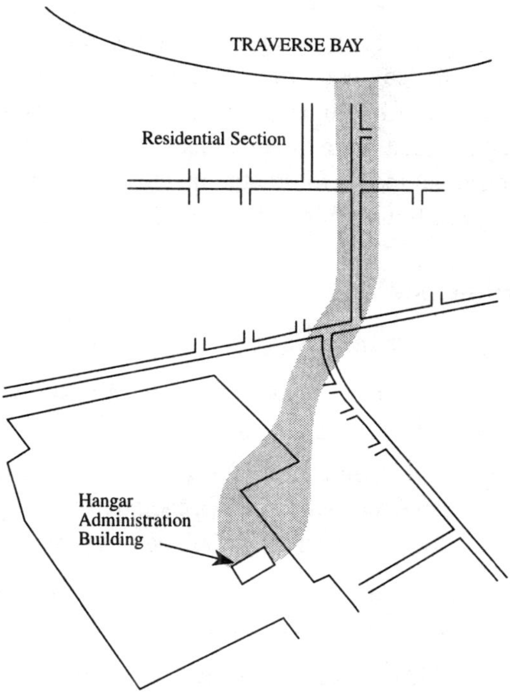

Ward et al. 1989

To determine the concentration of hydrocarbons in the experimental plot, core material was collected before implementing the liquid delivery system. Samples were collected from boreholes located along transects, which were parallel to groundwater flow, upgradient of the spill, and in the zone of heavy contamination. Three contiguous 1 m (3 ft) long cores from each borehole were collected to provide samples that extended:

- from the top of the capillary fringe to the lowest depth reached by the water table during an annual cycle;

- through the heavily contaminated zone; and
- from the slightly contaminated to the uncontaminated zone beneath the plume. The concentration of total extractable hydrocarbons was determined using gas chromatography (Vandegrift and Kampbell 1988). Using the data on total extractable hydrocarbons, the computer model BIOPLUME II was used to predict the amount of hydrogen peroxide that would be required to treat the contamination in the treatment plot. BIOPLUME II is a two-dimensional model for contaminant transport that is influenced by oxygen-limited biodegradation (Rifai et al. 1988).

Treatment Schedule

The treatment schedule included (Ward et al. 1989):

1) February 25, 1988 - injection of water to equilibrate the system hydraulically
2) March 1, 1988 - injection of water amended with pure oxygen and inorganic nutrients to adapt the microflora to greater-than-ambient concentrations of dissolved oxygen
3) June 1, 1988 - injection of water amended with 50 ppm hydrogen peroxide and inorganic nutrients
4) June 8, 1988 - injection of water amended with 100 ppm hydrogen peroxide and inorganic nutrients
5) June 15, 1988 - injection of water amended with 250 ppm hydrogen peroxide and inorganic nutrients
6) August 18, 1988 - injection of water amended with 500 ppm hydrogen peroxide and inorganic nutrients
7) December 1988 - injection of water amended with 750 ppm hydrogen peroxide and inorganic nutrients

Core and Groundwater Sampling

Core material and groundwater were collected periodically to determine the progress of the demonstration. During the demonstration, dissolved oxygen (DO), chloride (Cl), ammonia (NH_3), phosphate (PO_4), benzene, toluene, ethylbenzene, the xylene isomers (BTEX), pH, conductivity, and the water level were determined in samples of well water. Concentrations of DO, Cl, NH_3, and PO_4 were measured using standard methods (Standard Methods for the Examination of Water and Wastewater 1985). Concentrations of BTEX were measured using a modification of US EPA method 602 (Federal Register 1984) in which headspace analysis was used instead of purge-and-trap.

Microbial Numbers in Samples of Groundwater and Sediment

Viable counts of microoorganisms in groundwater and sediments were determined periodically during the demonstration (see previous section, Core and Ground Water Sampling) using the spread plate technique (Ward et al. 1989). Total heterotrophs were plated on Nutrient agar (Difco Industries, Detroit, MI) and hydrocarbon-degrading microorganisms were plated on 1.5% Noble agar (Difco Industries), incubated in the presence of aviation fuel vapors.

Treatability Study

The nutrient requirements of the microflora were determined by adding different combinations of inorganic nutrients to a mixture of 50 mL of groundwater, 2.0 ml of a 1:10 dilution of sediment, and 0.5 mL of aviation fuel (Ward et al. 1989). Combinations of KH_2PO_4 + Na_2HPO_4, with NH_4Cl, Na_2CO_3, $CaCl_2$, $MgSO_4$, and $FeSO_4$ were tested. Two treatments received only nitrogen or phosphorus. In addition, controls were prepared which received 1) groundwater and inoculum, 2) groundwater, inoculum, and gasoline, 3) groundwater and all inorganic nutrients, and 4) groundwater, all inorganic nutrients, and gasoline. The mixtures were incubated at room temperature (~25°C [77°F]) for 2 weeks, during which time gasoline was added every 2 to 3 days as needed. After incubation, the biomass was harvested by centrifugation, dried, and weighed.

System Design Using BIOPLUME II

The design parameters for the demonstration were selected using BIOPLUME II, and included the injection flow rate, number of injection wells, concentration of injected oxygen, and time required for remediation.

System Layout. The well system is diagrammed in figure A.2 (on page A.8). In December 1987, nine 10 cm (4 in) monitoring wells with 4 m (12 ft) long well screens and 12 small cluster wells were installed. The

Figure A.2
Schematic Diagram of Well System

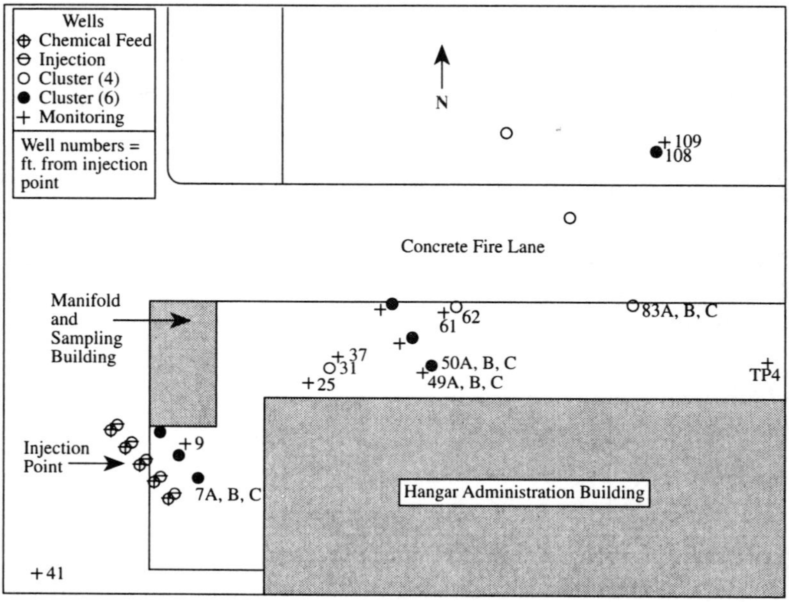

Ward et al. 1988

cluster wells had four, five, or six screens at different depths depending on their distance from the injection wells. The elevations of the screens were based on historical maximum and minimum water levels in a monitoring well (TP4) in the demonstration area. Cluster wells with four screens had sampling intervals at 4.3, 5.5, 6.0, and 6.4 m (14.0, 18.0, 19.5, and 21.0 ft) below the ground surface. Cluster wells with five and six screens had sampling intervals at 4.3, 5.5, 6.0, 6.4, and 6.9 m (14.0, 16.5, 18.0, 19.5, 21.0 and 22.5 ft). The sampling depths were numbered from 1 to 6, ranging from the shallowest to deepest screens. The cluster wells were constructed of stainless steel and extended into the manifold and sampling building. This building housed the manifold, sampling equipment, nutrients, and pure oxygen or peroxide.

Water was injected at a rate that would increase the water table by (0.3 m) (1 ft) to include the contaminated capillary fringe in the treatment zone and deliver enough oxygen for biodegradation in the contaminated interval (Ward et al. 1989). From a series of initial runs with BIOPLUME II, a flow rate of 150 L/min (40 gal/min) was selected. The injection water was pumped from a well located 180 m (600 ft) southeast of the test site and was split into two injection zones. Of the total flow, 110 L/min (29 gal/min) was pumped into five injection wells screened below the fuel spill between 7.7 to 8.8 m (25 to 29 ft) to raise the water table. The remaining 40 L/min (11 gal/min) was amended with nutrients and oxygen and pumped into chemical feed wells screened in the contaminated interval between 4.3 and 5.8m (14 and 19 ft).

The highest concentration of extractable hydrocarbon detected, 5,590 mg/kg wet solids, was used to calculate the total amount of contamination in the test plot; uniform distribution was assumed. The oxygen demand of the contamination was estimated using a oxygen or hydrogen peroxide:hydrocarbon ratio of 3.2:1 or 6.4:1, respectively. The model was used to determine the amount of oxygen required to bioremediate the test plot in 3 to 6 months using weekly, stepwise increases in oxygen concentration. The stepwise increase in oxygen concentration was used to allow the subsurface microorganisms to adapt to higher-than-ambient levels of oxygen.

The time needed for bioremediation of the test plot was estimated in a worst case scenario using BIOPLUME II. Initial runs with the model varied the retardation factor from 1 to 100. The retardation factor estimated the rate at which the sorbed and/or entrapped contaminants leached into the groundwater, with low numbers indicating little, and high numbers indicating a great deal of retardation. Although the use of different retardation factors results in different concentrations of dissolved contaminants, the total amount of contamination used in each scenario was the same. Therefore, the time required for bioremediation is dependent on the retardation factor used in the model. Using a retardation factor of 100 and an oxygen supply of 15,060 kg (33,890 lb), the worst-case scenario estimated that remediation of the test plot would require six months. This amount of oxygen was calculated assuming that hydrogen peroxide would be added at 8, 40, 100, 200, 400, and 800 mg/L during weeks 1, 2, 3, 4, 5, and 6, respectively, and then at 2,000 mg/L for the remaining six weeks. However, this design was not followed during the first 12 weeks of the demonstration.

System Design Using Microbial Treatability Study

Table A.1 (on page A.11) shows the nutrient amendments that resulted in more biomass than the control treatments, which received only nitrogen and phosphorus (alpha = 0.05). Although several treatments yielded more biomass than the nitrogen and phosphorus controls, the addition of NH_4Cl, KH_2PO_4, and Na_2HPO_4 was chosen as the simplest, least expensive, but effective amendment to enhance contaminant biodegradation.

System Operation

Water was injected into the test plot for 4 days to achieve hydraulic equilibrium, after which oxygen and inorganic nutrients were added. The nutrient amendment was prepared using food grade chemicals. Concentrated solutions of NH_4Cl, KH_2PO_4, and Na_2HPO_4 were prepared to yield 56,000, 28,000, and 28,000 mg/L, respectively, and mixed in a 1,900 L (500 gal) tank with injection water to produce groundwater feed concentrations of 381, 190, and 190 mg/L, respectively. The concentrations of chloride, phosphorus, amd ammonia nitrogen were 250, 75, and 100 mg/L, respectively. The injection water had a pH near neutrality and was 11° to 12° C (52° to 54°F). For each monitoring well, tracer tests were conducted to determine the actual seepage velocity of a chloride tracer and the inorganic nutrients along a flow path to that well (Wilson, Armstrong, and Rifai (1993).

Oxygen or hydrogen peroxide was added after the groundwater feed left the mixing tank. The average concentration of oxygen that was injected was approximately 40 mg/L.

Oxygen was injected into the aquifer for the first three months of operation, after which it was replaced with hydrogen peroxide. The hydrogen peroxide was added initially at 50 mg/L, and increased stepwise to a final concentration of 750 mg/L (see Core and Ground Water Sampling section for schedule). This schedule was different from that used by the BIOPLUME II model used to predict the time required for bioremediation of the test plot.

Table A.1
Mass of Cells Resulting From Different Inorganic Nutrient Amendments

Sample	KH_2PO_4 Na_2HPO_4	NH_4Cl	Na_2CO_3	$CaCl_2$	$MgSO_4$	$MnSO_4$	$FeSO_4$	Mass[a],g
20	+	+	-	-	+	-	-	0.0405
29	+	+	+	+	-	+	+	0.0405
16	+	+	-	+	+	-	-	0.0385
32	+	+	-	+	-	+	+	0.0380
9	+	+	-	-	+	-	-	0.0375
36	+	+	+	-	+	-	+	0.0350
14	+	+	+	-	-	+	-	0.0345
11	+	+	-	-	-		+	0.0335
33	+	+	-	+	+		+	0.0330
26	+	+	+	+	+	+	+	0.0330
13	+	+	+		+			0.0330
18	+	+		+			+	0.0325
15	+	+	+				+	0.0325
35	+	+	+			+	+	0.0320
38	+	+	+	+			+	0.0320
21	+	+				+	+	0.0320
31	+	+			+	+	+	0.0320
25	+	+	+	+	+	+	+	0.0315
39	+	+	+	+		+		0.0300
10	+	+				+		0.0290
30	+	+	+	+	+		+	0.0285
37	+	+	+		+	+		0.0265
23	+	+	+	+	+			0.0255
12	+	+	+	+				0.0250
7	+	+	+					0.0225
24	+	+	+	+	+	+		0.0225
27	+	+		+	+	+	+	0.0220
22	+	+	+	+				0.0215
8	+	+		+				0.0215
17	+	+		+		+		0.0210
34	+	+		+	+	+		0.0190
5	+	+						0.0185[b]
19	+	+			+	+		0.0180
3	+							0.0000
4		+						0.0000
28	+	+	+		+	+	+	0.0410

Fiorenza 1991

a Mean cell weight; a difference of 0.01593 g between treatments is significant
b Control treatment

Results and Discussion

Nutrient Transport in the Contaminated Interval

The seepage velocity of the injected water averaged 1.6 to 2.7 m (5 to 9 ft/day) (Wilson, Armstrong, and Rifai 1993). The results of tracer tests to determine the actual seepage velocities of the chloride tracer and inorganic nutrients are shown in table A.2. Transport of ammonium and and phosphate ions was about half that of water, a finding which is different from those in most other aquifers in which there is much stronger retardation of these ions. Within 2 months after injection of the nutrients, greater than 10 mg/L of ammonia nitrogen and phosphate were distributed throughout the test plot.

Oxygen Transport and BTEX Removal in Groundwater

The concentration of dissolved oxygen in groundwater collected from a well (BD-7B) 2 m (7 ft) from the injection point was similar to that in the chemical feed water (figures A.3 on page A.13, and A.4 on page A.14) (Wilson, Armstrong, and Rifai 1993). Although the concentration of dissolved oxygen exceeded the equilibrium solubility of oxygen at ambient

Table A.2

Seepage Velocity of Fronts of Injected Chloride, Oxygen, Ammonia, and Phosphate Between the Infiltration and Monitoring Wells

Distance from the infiltration wells (feet)	Level	Chloride	Oxygen	Ammonium ion	Phosphate
7B	2	10.0	5.6	6.5	5.6
(center well)	3	11.4	7.3	8.0	7.1
31	2	8.6	NBT	3.9	3.6
	3	8.0	4.4	3.9	3.7
50B	2	6.0	NBT	2.1	2.1
(center well)	3	5.5	NBT	3.3	2.9
62	2	4.2	1.2	2.6	1.7
	3	6.0	1.2	3.1	2.8
83B	2	2.5	NBT	1.9	0.7
(center well)	3	6.5	NBT	3.3	2.2

Wilson, Armstrong, and Rifai 1993

NBT = No breakthrough during the tracer test.

Figure A.3

Schedule of Supply of Oxygen or Hydrogen Peroxide to the Infiltration Wells

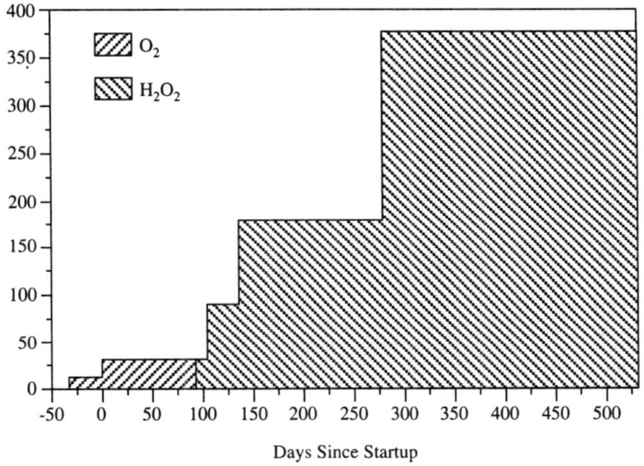

Data prior to day 0 is the ambient concentration of oxygen in the aquifer moving into the spill under the natural gradient. The concentration of hydrogen peroxide is expressed in terms of dioxygen equivalent.

Wilson, Armstrong, and Rifai 1993

hydrostatic presssure and temperature for most of the demonstration, these overpressures did not initiate bubble formation in the sand formation. The oxygen did not escape because there was no air space in the injection system.

Of interest were the patterns of oxygen and BTEX concentrations in groundwater collected from a well (BD-31) 9 m (31 ft) from the injection point (figure A.5 on page A.15) (Wilson, Armstrong, and Rifai (1993). The initial breakthrough of oxygen was at about the same concentration as that of the injection water (25 mg/L), after which it declined to nondetectable levels as the microflora acclimated to consume the oxygen. After 180 days oxygen broke through again, which corresponded to an increase in hydrogen peroxide concentration in the injection water (compare figures A.3 and A.5 on page A.15). After oxygen reappeared in the groundwater, BTEX soon disappeared.

Figure A.4

Breakthrough of Oxygen and Depletion of Total Alkylbenzenes in a Monitoring Well in the Most Contaminated Depth Interval Seven Feet from the Infiltration Wells

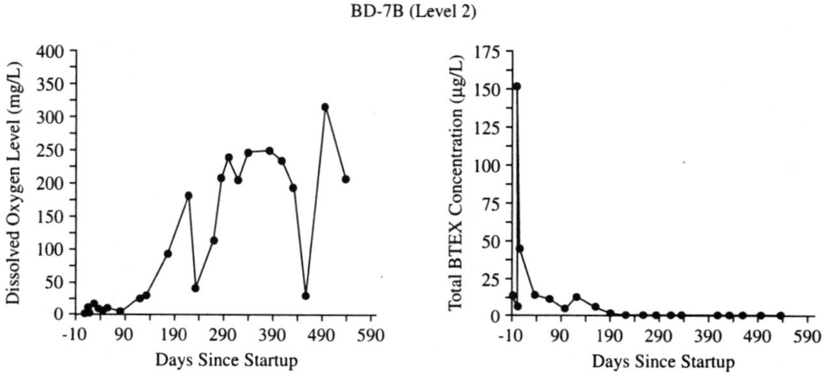

Wilson, Armstrong, and Rifai 1993

The dissolved oxygen concentrations in groundwater from wells located 15 m (50 ft) from the injection point were never above 10 mg/L (figure A.6 on page A.15) (Wilson, Armstrong, and Rifai 1993). Even though dissolved oxygen levels were low, BTEX was not detected after 300 days of treatment. Oxygen breakthrough was not detected at wells located 19 and 25 m (62 and 83 ft) from the injection point (figures A.7 on page A.16 and A.8 on page A.17). Although BTEX was still detectable in groundwater from these wells at the end of the demonstration, the concentrations were lower than initial levels.

Of interest was the apparent selective removal of benzene from the groundwater; the xylenes were more recalcitrant (Wilson, Armstrong, and Rifai 1993) (table A.3 on page A.18). The pattern of benzene removal was unusual. Benzene disappearance first occurred 25 m (83 ft) from the infiltration point below the contaminated zone; after 420 days of operation, removal was uniform throughout the plot. Benzene was removed near the infiltration wells using up to 80 pore volumes, whereas as little as eight pore

Figure A.5

Breakthrough of Oxygen and Depletion of Total Alkylbenzenes in a Monitoring Well in the Most Contaminated Depth Interval 31 Feet from the Infiltration Wells

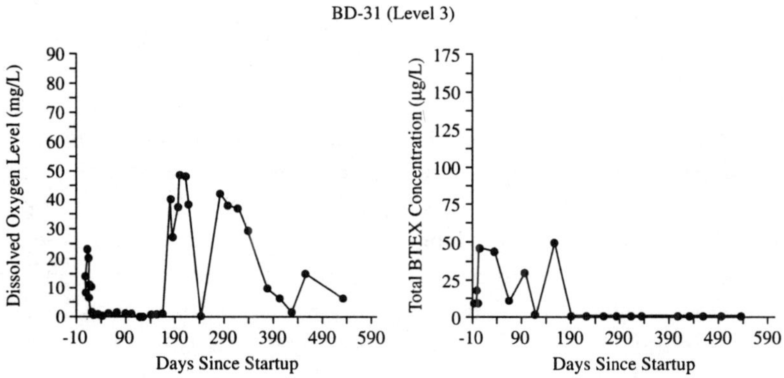

Wilson, Armstrong, and Rifai 1993

Figure A.6

Breakthrough of Oxygen and Depletion of Total Alkylbenzenes in a Monitoring Well in the Most Contaminated Depth Interval 50 Feet from the Infiltration Wells

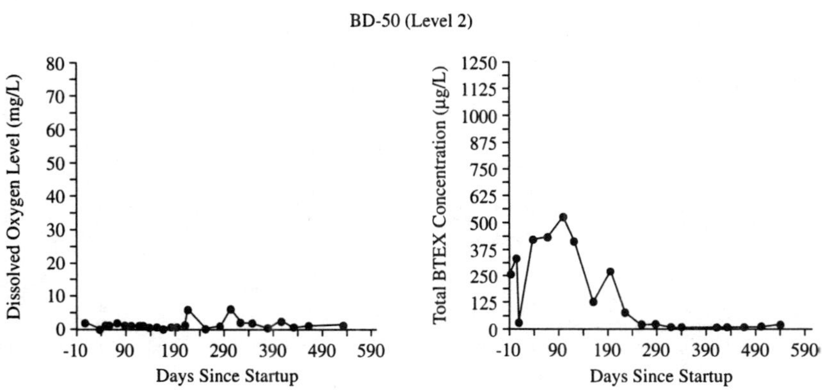

Wilson, Armstrong, and Rifai 1993

volumes was required to remove the aromatic compound 33 m (108 ft) from the infiltration wells.

Removal of BTEX and Petroleum Hydrocarbons from Sediments

Initial concentration of total petroleum hydrocarbon ranged from 2,000 to 12,000 mg/kg of which 5 to 10% was alkylbenzenes (Wilson, Armstrong, and Rifai 1993). Changes in concentrations of BTEX and total petroleum hydrocarbons are shown in table A.4 (on page A.19). Toluene and the xylenes were removed in samples collected near the monitoring well 10 m (32 ft) from the infiltration point (see sample 50T3). After 8 months of operation when oxygen broke through and BTEX disappeared in monitoring well BD-32, cores were collected and analyzed for BTEX and TPH. Concentrations of BTEX were below detection; however, concentrations of TPH remained at initial levels. After 12 months of operation, extensive removal of total petroleum hydrocarbons was detected (sample 50AQ3).

Figure A.7

Breakthrough of Oxygen and Depletion of Total Alkylbenzenes in a Monitoring Well in the Most Contaminated Depth Interval 62 Feet from the Infiltration Wells

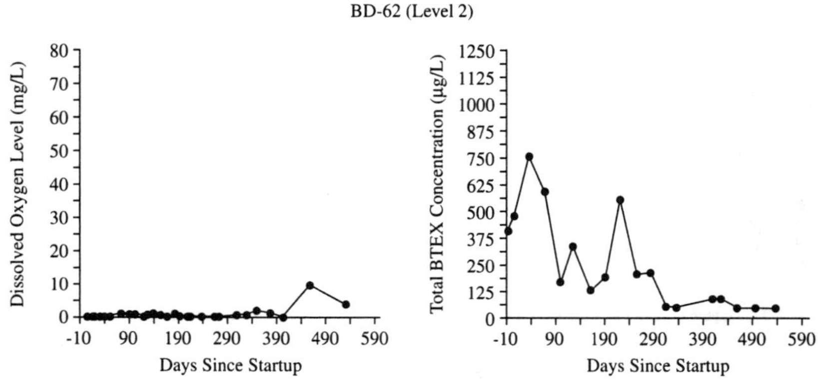

Wilson, Armstrong, and Rifai 1993

Figure A.8

Breakthrough of Oxygen and Depletion of Total Alkylbenzenes in a Monitoring Well in the Most Contaminated Depth Interval 83 Feet from the Infiltration Wells

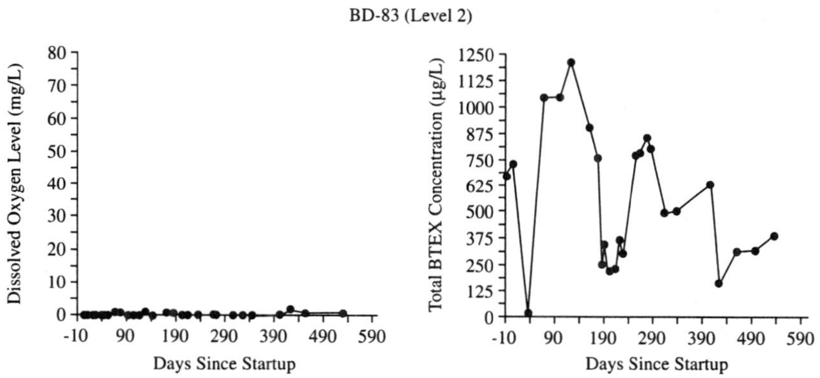

Wilson, Armstrong, and Rifai 1993

After 12 months of operation, toluene was the only alkylbenzene that had been removed completely from samples collected just in front of the monitoring well 19 m (62 ft) from the infiltration point; however, the concentration of TPH had been reduced from an initial concentration of about 6,500 to 3,100 mg/kg wet weight (table A.4 on page A.19) (Wilson, Armstrong, and Rifai 1993). This removal occurred even though high levels of oxygen had not broken through this section of the treatment zone (see figure A.7 on page A.16).

Concentrations of BTEX in samples of sediment at the end of the demonstration (522 days) were below 0.07 mg/kg in sediment samples collected 3, 10, and 16 m (10, 34, and 54 ft) from the infiltration point (table A.5) (Wilson, Armstrong, and Rifail 1993). In samples collected 19.5 m (64 ft) from the infiltration point, concentrations of benzene, toluene, and o-xylene were less than 0.07 mg/kg, whereas low concentrations of ethylbenzene and *m*-*p*-xylene persisted. Even though BTEX was extensively removed throughout the test plot, high concentrations of TPH remained.

Table A.3
Depletion of Alkylbenzenes in Groundwater After 17 Months of Infiltration of Nutrients and Oxygen

Distance from infiltration wells (meters feet)	Benzene	Toluene	Ethylbenzene	Total Xylenes
2.0 (7)	<0.1	<0.1	<0.1	<0.1
9.4 (31)	<0.1	<0.1	<0.1	0.8
15 (50)	<0.1	<0.1	1.0	10.4
19 (62)	<0.1	0.3	1.7	37.2
25 (83)	<0.1	0.3	12	367
33 (108)	<0.1	2.4	6.4	393

Wilson, Armstrong, and Rifai 1993

Microbial Patterns (adapted from Fiorenza, 1991)

The number of microorganisms in groundwater from monitoring wells did not follow any consistent pattern. Most likely, changes in microbial numbers were difficult to determine because these wells were screened over a 3 m (12-ft) interval where there were vast differences in hydrocarbon concentration. The number of heterotrophs or hydrocarbon-degrading organisms from a particular monitoring well approximated the average of all levels from the corresponding cluster well (see figure A.2 (on page A.8) for corresponding monitoring and cluster wells).

The data on cell numbers from cluster wells were more informative. The most dramatic and consistent changes in microbial numbers were detected in level 2 of the cluster wells, where the concentration of contamination was the highest. In samples with little or no contamination, there was no obvious trend.

As the concentration of BTEX began to decrease during the demonstration, dissolved oxygen increased. As the concentration of BTEX decreased in samples from the most contaminated interval (level 2 of the centerline cluster wells), the number of heterotrophs and hydrocarbon-degrading microorganisms in groundwater usually increased (figures A.9 and A.10 on page A.20, A.11 on page A.21, A.12 on page A.22, A.13 and A.14 on page A.23). This trend was not evident in wells located 25 and 33 m (83 and 108

Table A.4

Changes in Concentration of Alkylbenzenes and Total Petroleum Hydrocarbons in Core Material During Bioremediation of an Aquifer Contaminated With Aviation Gasoline

Core*	TPH	Benzene	Toluene	Ethyl-Benzene	o-Xylene	m+p Xylene
		------------------------ (mg/kg wet sample) ------------------------				
Near BD-31, collected June 1988 after 4 months of perfusion with mineral nutrients and oxygen						
50T3	3,330	1.4	<1	7.3	23	<1
Near BD-31, collected after 8 months of perfusion with mineral nutrients and oxygen						
50AE4	8,400	<0.3	<0.3	<0.3	<0.3	<0.3
50AE5	2,370	<0.3	<0.3	<0.3	<0.3	<0.3
Near BD-31, collected after 12 months of perfusion with mineral nutrients and oxygen						
50AQ3	9	<0.3	<0.3	<0.3	<0.3	<0.3
Near BD-62, collected after 12 months of perfusion with mineral nutrients and oxygen						
50AR4	3,100	1.5	<0.3	9.2	30	6.2

Wilson, Armstrong, and Rifai 1993

*Number designates distance from infiltration gallery, letters indicate sample.

ft) away from the infiltration point where groundwater contamination still existed at the end of the demonstration. Also of interest was the decrease in both heterotrophs and hydrocarbon-degrading microorganisms to nondetectable levels in samples from well BD-7B after the BTEX disappeared (figure A.9 on page A.20). This plummet in numbers coincided with the detection of hydrogen peroxide in the groundwater from this well, and suggests toxicity of the peroxide to the microorganisms. However, low numbers of microorganisms (<100 cells/mL) were detected in samples from this well later in the demonstration while peroxide concentrations were still high, suggesting recolonization by peroxide-tolerant bacteria.

The number of catalase-postive microorganisms increased in samples of groundwater from all levels of the cluster wells as the demonstration proceeded (data not shown). Catalase is the enzyme that catalyzes the breakdown of hydrogen peroxide to oxygen and water. A deviation from this trend was seen in samples from well BD-7B in which the number of cata-

Figure A.9

Effect of DO (●) and BTEX (×) on the Number of Heterotrophs (■) and Hydrocarbon-Degraders (□) in Groundwater from BD-7B, Level 2

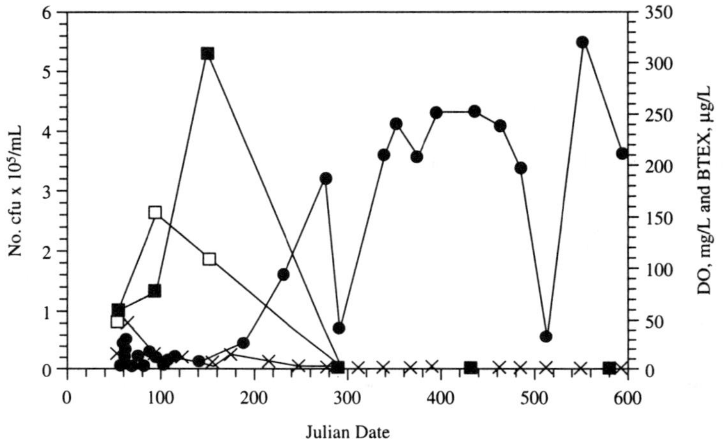

Fiorenza 1991

Figure A.10

Effect of DO (●) and BTEX (×) on the Number of Heterotrophs (■) and Hydrocarbon-Degraders (□) in Groundwater from BD-31, Level 2

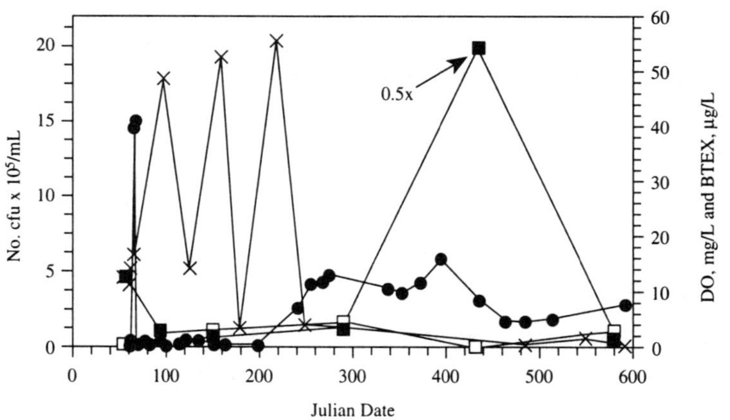

Fiorenza 1991

Figure A.11

Effect of DO (●) and BTEX (×) on the Number of Heterotrophs (■) and Hydrocarbon-Degraders (□) in Groundwater from BD-50, Level 2

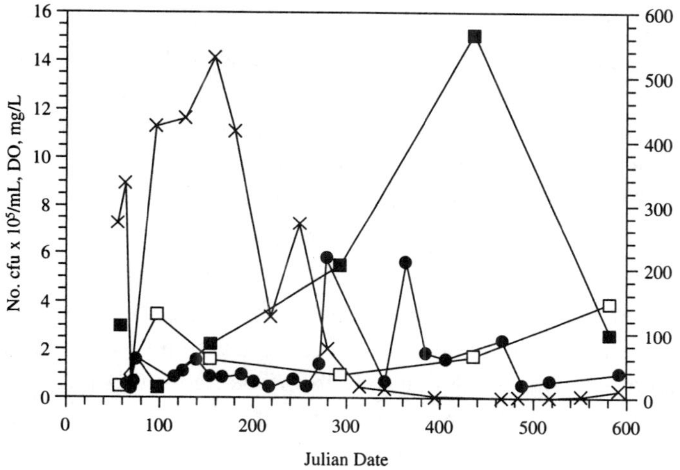

Fiorenza 1991

lase-positive organisms increased until the BTEX disappeared and peroxide was detected (described earlier); then there were no detectable microorganisms. At the end of the demonstration, the percentage of microoganisms that were catalase-positive was greater than 90% in most groundwater samples. These data suggest that catalase-positive organisms were selected by exposure to peroxide, and those lacking catalase were unable to withstand the addition of peroxide.

Samples of core material from the shallow, contaminated and the deep, uncontaminated zones also were analyzed for microbial numbers. The change in number of hydrocarbon-degrading and heterotrophic microorganisms was similar to that observed in groundwater. Counts were higher in the shallow, contaminated samples than the deep, uncontaminated samples. Both types of organisms increased in concentration in both shallow and deep samples collected 9.4 and 19 m (31 and 62 ft) from the injection point and then declined (figures A.15 on page A.24, A.16 on page A.25). In samples collected 33 m (108 ft) from the injection point, numbers of both types of organisms initially were low and declined during the demonstration (figure A.17 on page A.25).

For samples from the 9.4 m (31-ft) zone, organisms were not detected in the deep, uncontaminated samples and lower numbers than previously were detected in the shallow, contaminated samples by Julian date 580; this occurred about 400 days after BTEX was no longer detectable in the 9.4-m (31-ft) well near this sampling point. Because BTEX was no longer detectable and therefore carbon was not available for cell metabolism and catalase production, hydrogen peroxide may have sterilized the area.

The change in catalase activity in shallow and deep cores is shown in figure A.18 (on page A.27). Except for samples collected close to the infiltration point (2 and 9.4 m (7 and 31 ft)), catalase activity increased during the demonstration, suggesting that the subsurface microflora adapted to the peroxide addition by increasing catalase activity. The decline in activity in

Figure A.12

Effect of DO (●) and BTEX (×) on the Number of Heterotrophs (■) and Hydrocarbon-Degraders (□) in Groundwater from BD-62, Level 2

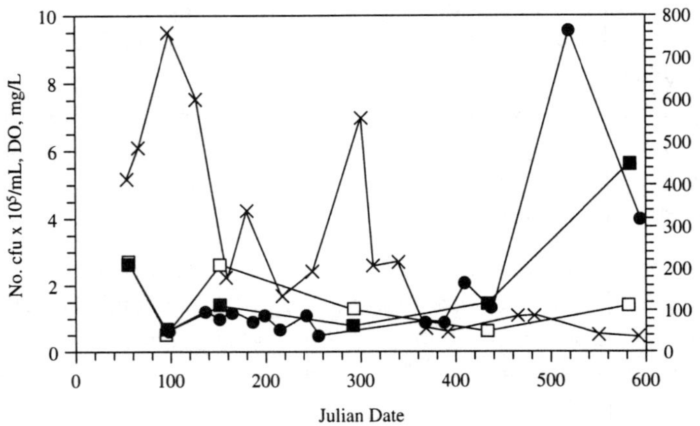

Fiorenza 1991

Figure A.13

Effect of DO (●) and BTEX (×) on the Number of Heterotrophs (■) and Hydrocarbon-Degraders (□) in Groundwater from BD-83B, Level 2

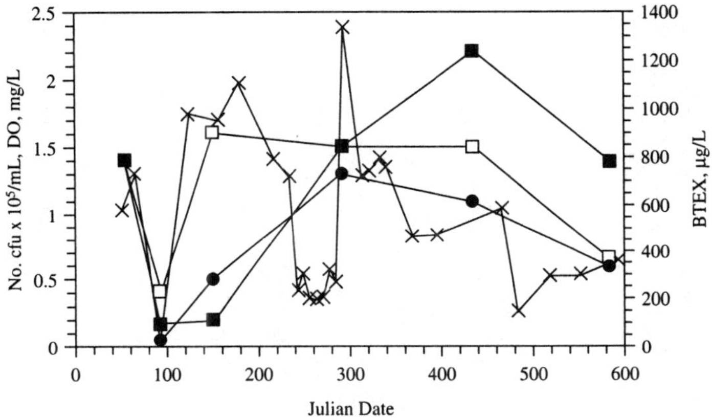

Fiorenza 1991

Figure A.14

Effect of DO (●) and BTEX (×) on the Number of Heterotrophs (■) and Hydrocarbon-Degraders (□) in Groundwater from BD-108, Level 2

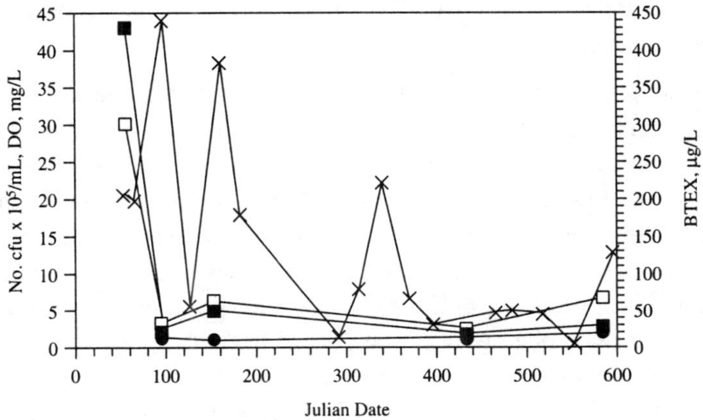

Fiorenza 1991

Figure A.15

Heterotrophs (Squares) and Hydrocarbon-Degraders (Circles) in Shallow (Closed Symbol) and Deep Subsurface (Open Symbol) Cores at 31 Feet

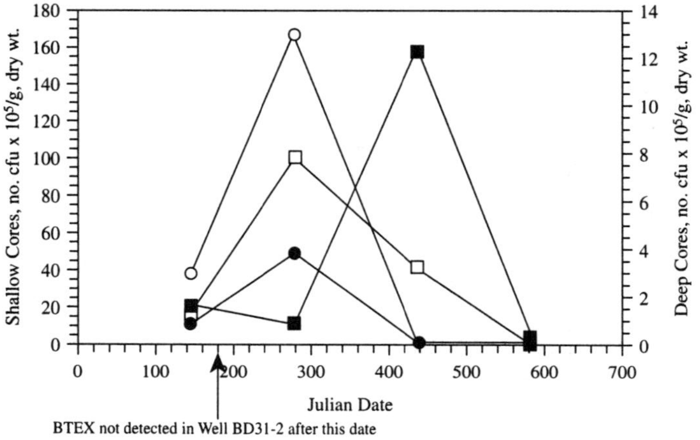

Fiorenza 1991

samples collected close to the infiltration point may have resulted from the lack of a carbon source (BTEX) for microbial utilization during the latter phase of the treatment and, therefore the inability of the microorganisms to produce catalase.

Catalase activity was higher in the shallow, contaminated samples, which contained BTEX that could be used as a carbon source, than in the deep, uncontaminated samples (figure A.18 on page A.26). The decreases in activity in shallow samples observed at Julian day 279 occurred after a decline in the water table elevation that prevented transport of nutrients and oxygen through this zone; declines in heterotrophs were observed in shallow samples collected 9.4 m (31 ft) from the injection point (compare figures A.15, A.18 on page A.26). The decline in catalase activity in deep samples collected 9.4 m (31 ft) from the injection point may have resulted from toxicity of peroxide, which was detected at 167 mg/L at level 4 after 533 days of operation (compare table A.5 on page A.27 and figure A.18 on page A.26).

Figure A.16

Heterotrophs (Squares) and Hydrocarbon-Degraders (Circles) in Shallow (Closed Symbol) and Deep Subsurface (Open Symbol) Cores at 62 Feet

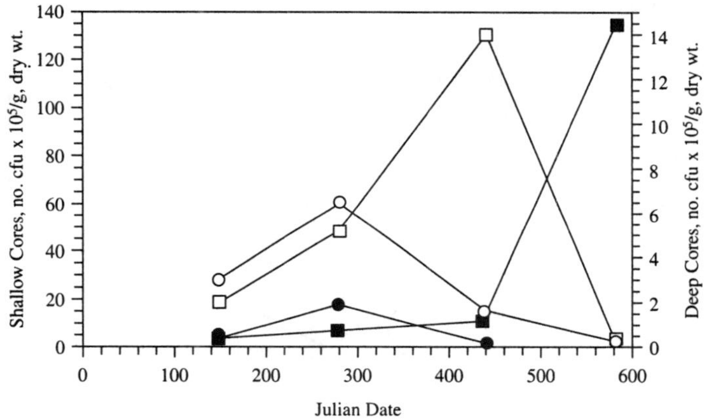

Fiorenza 1991

Figure A.17

Heterotrophs (Squares) and Hydrocarbon-Degraders (Circles) in Shallow (Closed Symbol) and Deep Subsurface (Open Symbol) Cores at 108 Feet

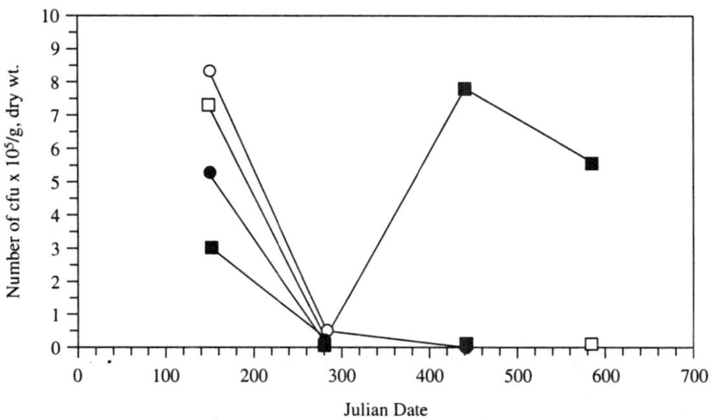

Fiorenza 1991

Figure A.18
Catalase Activity in Shallow and Deep Subsurface Cores 7 (◆), 31 (■), 62 (●), and 108 (▲) ft from the Infiltration Wells

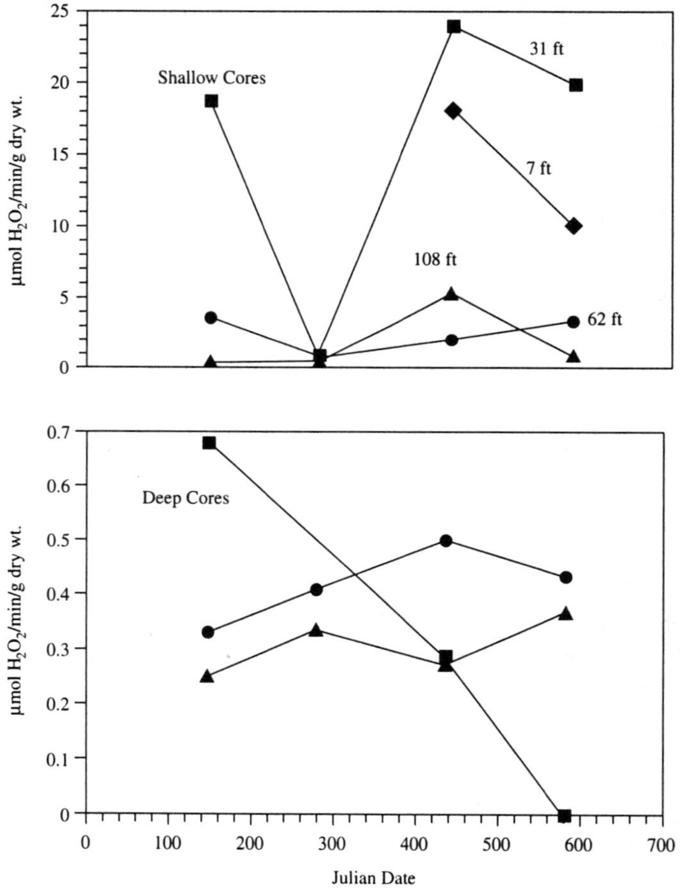

Fiorenza 1991

Nitrate and Alkylbenzene Biodegradation

Although high levels of oxygen were never detected in wells beyond 15 m (50 ft) from the infiltration point, significant amounts of alkylbenzenes were removed in this region. Anaerobic biodegradation of these compounds was believed to be the result of using nitrate as an alternate electron acceptor (Wilson, Armstrong, and Rifai 1993). The biodegradability of the alkylbenzenes, except benzene, using nitrate as the terminal electron acceptor, has been demonstrated in the laboratory and field. The remaining straight-chain fraction is recalcitrant under anaerobic conditions.

Although nitrate was not infiltrated into the test plot, nitrate accumulated in zones that contained oxygen, little or no hydrogen peroxide (table A.6 on page A.28), and no BTEX (figures A.4 to A.8 on pages A.14 to A.17); however, nitrate levels were lower in zones downgradient that still contained BTEX. Calculations indicated that a significant fraction of the oxygen added as hydrogen peroxide was converted to nitrate. These data suggest that the nitrate produced from nitrification of the added ammonia served as the terminal electron acceptor in BTEX biodegradation in zones that received little or no oxygen.

Effectiveness of Treatment

After 17 months of treatment, removal of BTEX from groundwater was extensive, whereas that for the TPH was low (Wilson, Armstrong, and Rifai

Table A.5

Concentrations of Alkylbenzenes and Total Petroleum Hydrocarbons in Core Material From Most Contaminated Interval, After 17 Months of Perfusion with Mineral Nutrients and Oxygen

Core	TPH	Benzene	Toluene	Ethyl-Benzene	o-Xylene	m+p Xylene
	---------- (mg/kg wet sample) ----------					
Near BD-7, 3 m (10 ft) from the infiltration wells						
50AY3	922	<0.07	<0.07	<0.07	<0.07	<0.07
Near BD-31, 10 m (32 ft) from the infiltration wells						
50BD2	2,310	<0.07	<0.07	<0.07	<0.07	<0.07
Near BD-50B, 16 m (53 ft) from the infiltration wells						
50AW42	10,800	<0.07	<0.07	<0.07	<0.07	<0.07
Near BD-62, 19.5 (64 ft) from the infiltration wells						
50BB7	1,280	<0.07	<0.07	3.2	<0.07	0.14

Wilson, Armstrong, and Rifai 1993

Table A.6

Hydrogen Peroxide Decomposition, Nitrification, and Potential Denitrification After Infiltration With Mineral Nutrients and Hydrogen Peroxide

Distance (feet)	Level	O_2	H_2O_2	NH_4+	NO_3-	NO_2-	Denitrified
	--------mg/liter--------			--------------mg N/liter-----------			
Injection				106	0.30	<0.05	
7B (center well)	2	210	930	91	0.35	<0.05	15
	3	212	930	93	0.30	<0.05	13
	4	230	840	92	0.35	<0.05	14
	5	150	860	95	0.40	<0.05	11
	6	1.9	0.5				
31	2	8.2	<0.1	68	30	1.9	6
	3	6.5	<0.1	34	65	2.4	5
	4	137	167	82	4.3	1.9	18
50B (center well)	1	3.7	0.1	67	17	10	12
	2	1.0	<0.1	64	22	2.8	17
	3	0.4	<0.1	17	80	1.9	7
	4	1.0	<0.1	72	20	2.4	12
62	2	4.0	0.2	58	12	12	36
	3	0.7	<0.1	59	27	1.4	19
	4	0.9	<0.1	42	57	3.2	4
83B (center well)	2	0.6	<0.1	39	18	2.2	47
	3	1.1	0.2	60	19	7.0	20
	4	1.2	0.1	18	42	4.4	42
108	2	0.4	<0.1	1.8	3.2	<0.05	101
	3	1.6	<0.1	13	4.3	7.3	81
	4	0.7	<0.1	19	15	0.9	71
	5	1.0	<0.1	19		1.0	71

Wilson, Armstrong, and Rifai 1993

Hydrogen peroxide and oxygen data collected, 8/16/89 after 533 days of operation. Ammonia, nitrate, and nitrite data collected 7/13/89 after 499 days of operation.

1993). Total remediation of the test plot was not expected since the amount of oxygen added was less than that required to achieve complete biodegradation of the hydrocarbons.

At 34 m (110 ft) from the infiltration point, the concentration of benzene was less than 1 mg/L and the concentrations of the other alkylbenzenes were below federal drinking water standards. The concentration of each alkylbenzene in samples of core material collected up to 15 m (50 ft) from the infiltration point was less than 0.07 mg/kg. However, the concentration of TPH in the most contaminated interval ranged from 992 to 10,800 mg/

kg. Based on the amounts of oxygen and nutrients added to the treatment zone, the selective removal of the alkylbenzene fraction is difficult to explain. However, a significant amount of the oxygen supplied as hydrogen peroxide was converted to nitrate. The overall selective removal of the BTEX fraction probably resulted from the use of nitrate as the terminal electron acceptor in regions that received little or no oxygen.

IN SITU BIOREMEDIATION OF DIESEL FUEL IN SOIL Maureen E. Leavitt, IT Corporation, Knoxville TN

The largest inland oil spill to date occurred in 1988 in western Pennsylvania. A tank collapse at a petroleum terminal resulted in the release of a million gallons of No. 2 fuel oil. After the emergency response was completed, an estimated 562,000 L (145,000 gal) of oil remained on the soils surrounding the tank area and in a nearby wooded area. The IT Corporation was contracted to investigate, design, and implement a bioremediation strategy to reduce the TPH concentration in soil. The proposed approach was to stimulate the indigenous hydrocarbon-degrading bacteria using inorganic nutrients and regular tilling. Since the spill was recent, the contamination was limited to the top two feet. Therefore, it was decided that all soils would be treated in place.

This project was among the first bioremediation applications for this region of the Pennsylvania Department of Environmental Resources, and proof of the treatment was required prior to initiating the full-scale application. Samples were collected within two months of the spill. A bioassessment was completed to confirm that conditions were conducive to bioremediation across the site. In addition, a biotreatability study was conducted to demonstrate that indigenous organisms could reduce the TPH content when supplied with nutrients, lime, and tilling.

Full-scale operations began within six months of the release. The soils monitoring program documented a substantial increase in the bacterial populations density, and maintenance of pH within the desirable range.

Leachate quality improved during the bioremediation program, reaching levels acceptable to the existing National Pollutant Discharge Elimination System (NPDES) permit and the Consent Decree.

Although excessive rainfall often inhibited effective tilling, the 1,000 mg/kg target level was achieved in many areas after two growing seasons. The third and final treatment season resulted in attaining the assigned treatment target level in all areas, with the exception of one isolated area which was identified as having non-spill related hydrocarbons. Extensive closure sampling and analyses were completed and the soils treatment program was considered a success.

Introduction

Land treatment (or land farming) for the reduction of petroleum hydrocarbons in soil has been applied throughout the petrochemical industry for many years. The conventional system utilizes a multi-acre field over which the waste is applied. The field is regularly tilled to oxygenate the soils, to redistribute the waste throughout the soil medium, and to supplement the soil with moisture and nutrients (Bartha and Bossert 1984). Successful land treatments propagate an adapted hydrocarbon-degrading bacterial population that is capable of mineralizing complex hydrocarbon loadings in excess of 2% by weight.

The described adaptation of land treatment was used to remediate soils that have been unintentionally contaminated with fuel oil. This method of land treatment did not have the benefit of an even, calculated loading rate, nor of an ideal soil matrix. Several physical, geographical, and regulatory issues were addressed and resolved, resulting in successful remediation of the subject soil.

Pertinent Site Factors

The subject fuel was No. 2 or diesel fuel, which is generally between 1% to 10% volatiles (Reed and Associates, Inc. 1988). Although almost 16 million liters (4 MMgal) of fuel was stored in the tank, estimates stated that approximately 4 million liters (1 MMgal) was released. After emergency response efforts were completed, approximately 562,000 L (145,000 gal) remained on the soil. Total petroleum hydrocarbon concentrations were as high as 100,000 mg/kg; however, most of the soil contained concentrations in the 10,000 to 20,000 mg/kg range.

The area impacted by the fuel release was an operating petroleum terminal (figure A.19). The impacted area included several tank basins, an asphalt plant area, and a wooded/field area across a highway from the terminal. The total volume of soil to be treated was approximately 11,500 m^3 (15,000 yd^3). The topography was generally flat and sloped toward the river. The soils that received the fuel consisted of clays to clay/silts. The ambient air temperature at the time of the release was -4°C (25°F).

Remedial Approach

Samples were collected from across the site to determine the potential for successful bioremediation. A summary of the results is listed in table A.7 (on page A.32). A total of 27 samples were collected and analyzed for pH, background nutrient concentration, and microbial population density. The

Figure A.19
Site Configuration

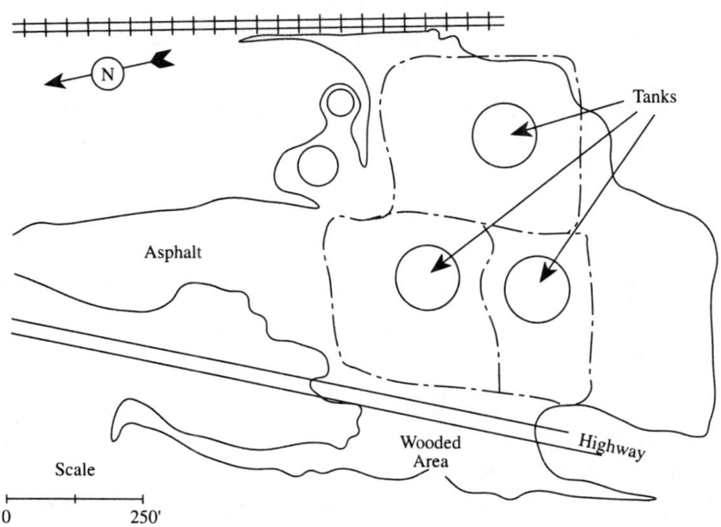

Table A.7
Initial Bioassessment of Site Soils

Area	pH	Ammonia (mg/kg)	Phosphate (mg/kg)	Total Heterotrophs (cfu/g)	Diesel Degraders (cfu/g)
Asphalt Plant[A]	6.9	8	0.5	2.6 x 106	2.5 x 106
Tank Basins[B]	5.2	8	0.5	6.9 x 105	7.8 x 105
Wooded Area[C]	5.5	12	47	1.1 x 107	1.5 x 106

A = Average of 7 samples
B = Average of 10 samples
C = Average of 10 samples
cfu = colony-forming unit

pH was near-neutral in the asphalt plant area; however, the tank basin and wooded areas exhibited acidic pH values. The background nutrient concentrations were negligible. The microbial population density ranged from 10^4 to 10^7 colony-forming units per gram, for both heterotrophs and hydrocarbon-degrading bacteria. These data suggested that a significant microbial population existed, although nutrient and lime augmentation would be required to stimulate rapid biodegradation of the diesel fuel.

To determine the specific requirements for the full-scale system, and to establish that biodegradation could reduce the TPH content of the soil, a biotreatability study was completed and has been described (Leavitt and Jadlocki 1989). The study evaluated the benefit of lime addition, and four different fertilizer loadings compared to moisture addition and tilling alone. The specific treatments are described in table A.8 (on page A.33). The treatments were maintained by maintaining the target moisture content (approximately 17% by weight) and thorough weekly mixing. The prescribed nutrient dosage for each treatment was added in three aliquots, one every two weeks. At two week intervals, duplicate samples from each treatment were collected and analyzed for TPH.

The results of the TPH analyses are illustrated in figure A.20 (on page A.34). There was no discernible difference between the nutrient dosages, and the TPH in each nutrient treatment was considerably lower than the untreated controls. In addition, the bacterial densities increased by as much as two orders of magnitude in the nutrient-amended treatments. The greatest increase in bacterial density was observed in the manure treated soil. Comparing the effect of lime addition, higher bacterial densities were observed in soil samples after lime treatment. The results of the study suggested that lime and nutrient addition increased the microbial population, resulting in a lower TPH content compared to the control (moisture and tilling only). These results were used to design a full-scale operating plan.

The initial proposal suggested treating soils using nutrient and lime addition and tilling to reach a 5,000 mg/kg target level over two growing seasons. The final Consent Decree ordered treatment of the soils as described to 1,000 ppm over three growing seasons.

Since the soils were to be treated in place, random sampling was not recommended due to the potential for high variability in contaminant concentrations. Instead, specific areas were chosen to be monitored once every three weeks to determine the nutrient and TPH content, as well as the microbial density and pH. Samples were collected at three distinct depths for TPH analysis to identify any vertical migration of contamination and to confirm that complete mixing was occurring.

In addition to soil, water that accumulated in several lysimeters placed within the unsaturated zone were sampled to document any contaminant migration through leachate resulting from the treatment program. Samples

Table A.8

Bench-Scale Biotreatability Study Treatment Scheme

Microcosm Number	Water	Lime	Fertilizer Content (N:P:K)	Target Nutrient Ratios (C:N:P)[a]
1	Yes	No	None	None
2	Yes	Yes	None	None
3	Yes	No	2:0.8:2[b]	100:10:1
4	Yes	Yes	20:4:10	100:10:1
5	Yes	Yes	10:6:4	100:3:1
6	Yes	Yes	20:4:10	100:3:0.3
7	Yes	No	20:4:10	100:10:1
8	Yes	No	10:6:4	100:3:1

a If C:N:P ratios could not be met exactly, C:N was chosen as the priority ratio and nitrogen loading was used to determine the quantity of fertilizer required.
b Fertilizer was processed cow manure.

Figure A.20
Biotreatability Study Total Petroleum Hydrocarbon Analysis

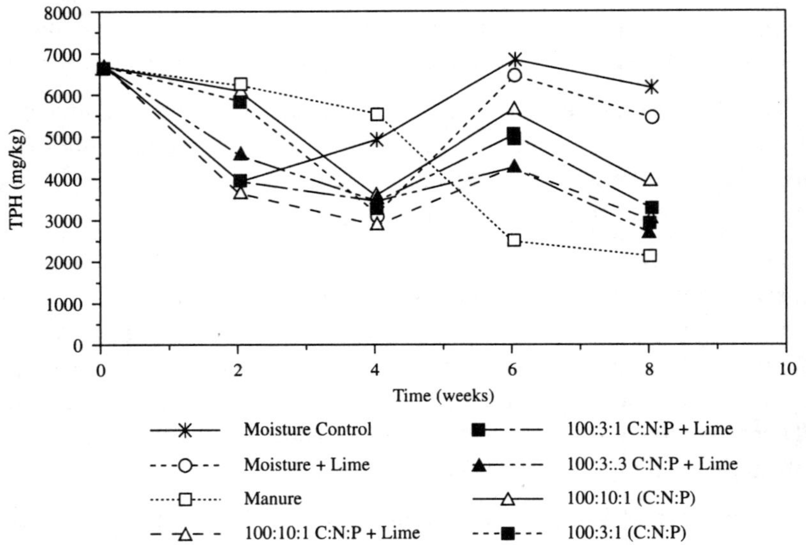

collected from these lysimeters, as well as soils, were tested for volatile and semivolatile contaminants including benzene, naphthalene, 2-methylnaphthalene, toluene, xylenes, ethylbenzene and phenanthrene.

Field Progress

The first growing season began in July of 1988 and continued through November of 1988 (table A.9 on page A.35). During this 105-day period, there were 26 days of rain. The moisture accumulation resulting from the rain limited the ability to mix soils and hampered oxygen diffusion into the soil. Nutrient addition was applied in monthly doses to maintain a 100:10:1 C:N:P ratio (approximately 200 pounds per acre per application). The pH was maintained between 6.0 and 7.6 in all areas using lime addition at a rate of 4,000 pounds per acre.

Table A.9
Bioremediation Program Operating Record

Growing Season	Operating Months	Total Operating Days	Days Of Rain	Percent Rain Days
1988	Jul - Nov	105	26	25
1989	Apr - Oct	199	82	41
1990	May - Oct	190	73	38

The microbial density of both heterotrophs and diesel-degrading bacteria increased as much as three orders of magnitude, providing a sufficient microbial density for biodegradation (data not shown).

At the completion of the first season, an overall average TPH reduction between 44 and 81 percent was observed (table A.10). Redidual TPH levels ranged from 2,000 mg/kg to 8,000 mg/kg.

Analysis of leachate collected from lysimeters is shown for the beginning and end of the first two treatment periods (table A.11a and b on page A.36). Each value represents the average of between nine and fifteen samples. In many samples, both volatile and semivolatile contaminants were below the detection limits. Over the two treatment periods, an in-

Table A.10
Summary of Soil Analytical Results — 1988 Season

Area	March 1988	Autumn 1988	Percent Reduction
	---- TPH concentration, mg/kg ----		
Basin A	8,844	4,588	48
Basin B	17,569	7,823	55
Basin C	4,200	2,340	44
Asphalt	11,980	6,737	44
Wooded	10,801	2,026	81

Table A.11a

Analysis of Leachate Collected in Lysimeters Over Two Growing Seasons

Date	Benzene	Ethylbenzene	Toluene	Xylenes
	µg/L			
July 1988	8	28	23	144
November 1988	4	ND (5)	ND (5)	ND (5)
April 1989	ND (5)	ND (5)	ND (5)	ND (5)
September 1989	9	12	8	29

Table A.11b

Analysis of Leachate Collected in Lysimeters Over Two Growing Seasons

Date	Methyl-Napthalene	Naphthalene	Phenanthrene
	µg/L		
July 1988	40	61	ND (10)
November 1988	ND (10)	ND (10)	ND (10)
April 1989	ND (10)	ND (10)	ND (10)
September 1989	ND (10)	15	ND (10)

crease in contaminant concentrations was not detected. Therefore, it was concluded that the treatment program did not cause a deterioration in the leachate quality.

The second growing season began in April of 1989 and continued through October of 1989. In this 199 day period, 82 days of rain were documented (table A.9 on page A.35). The excessive moisture resulted in severe tilling limitations. However, it was established that the stratified sampling could be discontinued after statistical analysis proved the soils to be vertically homogenized. The bacterial population densities were as high as 10^{12} cfu/g, with an average between 10^9 and 10^{10} cfu/g.

The concentrations of volatile and semivolatile compounds, in samples collected from lysimeter continued to be below detection, therefore lysimeter sampling was discontinued. Average TPH values at the completion of this season varied from 1,600 mg/kg to 6,500 mg/kg (table A.12). Isolated areas achieved the 1,000 mg/kg target level.

The third growing season began in May of 1990 and continued through October, 1990. Severe rainfall hampered any activity prior to May. Approximately 73 days of rain occurred during the 190 day period. Also during that period, the bacterial density declined from an average of 10^9 to an average of 10^7 cfu/g. Since nutrient addition and tilling remained constant, the decline was presumed to be attributable to lack of available organic carbon for microbial maintenance and growth.

At the start of the third season, TPH values ranged between 2,000 and 7,000 mg/kg TPH. Initial analytical results indicated that most of the terminal soils were nearing closure as defined in the Consent Decree. Therefore,

Table A.12
Summary of Soil Analytical Results — 1989 Season

Sample Location	Average TPH (mg/kg)
Asphalt Area	3,791
Asphalt Area	6,368
Asphalt Area	6,238
Tank Basins	5,422
Tank Basins	4,358
Tank Basins	4,211
Tank Basins	4,967
Tank Basins	3,617
Tank Basins	2,822
Tank Basins	1,681
Tank Basins	3,771
Tank Basins	1,606
Wooded Area	3,933
Wooded Area	1,587
Wooded Area	1,889
Wooded Area	2,132
Wooded Area	3,831
Wooded Area	2,361

while tilling and nutrient addition proceeded, the terminal was divided into 779 grids of approximately 250 square feet, as directed for site closure. As the samples achieved a TPH value of less than 1,000 mg/kg, the corresponding grid was closed, and no further tilling or nutrient addition occurred. Most of the area achieved this criterion during the summer months.

Completion

A total of twenty of the 779 grids exhibited TPH values greater than 1,000 ppm at the end of 1990. Since these grids were treated exactly like the rest of the terminal, and since the program was successful in removing the diesel from the rest of the terminal, it was believed that the diesel had been completely removed from these grids as well. Further investigation into the nature of the remaining contaminants suggested that these areas may have been impacted by the asphalt plant operation and could be attributable to tank bottom sludges.

The final TPH values across the site averaged 450 mg/kg (table A.13). Most of the area was found to have values within 50 percent of this value. Two examples of actual sample location and variability are included in figures A.21 and A.22 (on page A.39). The soil bioremediation program was completed and considered successful in mitigating the petroleum hydrocarbon related to the January, 1988 spill.

Table A.13
Summary of Soil Analytical Results — 1990 Season

Area	Number Of Grids	Average TPH (mg/kg)	
		01/02/88	12/31/90
Basin A	73	10,851	377
Basin B	197	18,838	395
Basin C	233	10,851	439
Basin D	100	14,488	442
Wooded Area	119	8,516	379
Asphalt Area	57	15,143	928
Terminal Average	779	12,988	450

Figure A.21
Closure Sampling Record

Tank Basin
TPH (mg/kg)

<50	650	820	430	420	170	230	290	470		
610	430	720	870	930	530	830	720	540		
400	400	640	500	570	680	350	540	360	540	
<50	900	720	720	650	360	360	340	980	940	
740	500	710	600	430	320	210	310	310	270	740
530	460	780	460	430	430	180	320	300	340	770
700	330	800	600	500	150	190	79	78	180	110
810	370	600	120	160	150	78	71	92	130	
500	930	310	190	370	150	81	190	120	300	
640	570	470	640	310	190	160	170	180		

Figure A.22
Closure Sampling Record

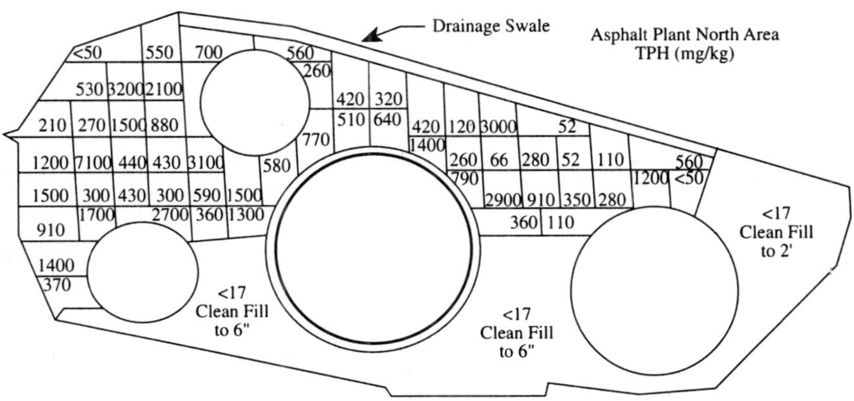

B

GLOSSARY OF TERMS

Abiotic Abiotic means nonbiological.

Aerobic Aerobic means in the presence of oxygen.

Anaerobic Anaerobic means in the absence of oxygen.

Anoxic Anoxic means in the absence of oxygen.

Aquifer An aquifer is an underground geologic formation that is filled with water and which is permeable enough to transmit water to wells and springs.

Arrhenius plot An equation relating the rate of a chemical reaction to temperature can be expressed by an Arrhenius plot; in general, reaction rates increase by a factor of two or three for each 10°C rise in temperature; departures from the Arrhenius plot are indicative of certain processes, e.g. reactions controlled by enzymes.

Bioaugmentation Bioaugmentation is the repeated addition of microbes that degrade but do not grow on the contaminants in a bioremedial process.

Biogeochemistry Biogeochemistry is the involvement of living organisms with the earth's chemistry. Biogeochemical processes control the global cycling of the biologically important elements carbon, nitrogen, phosphorus, and sulfur, as well as the cycling of a variety of trace elements.

Bioreactor Bioreactors are ex situ biodegradation treatment facilities; many types of bioreactors have been developed for various treatment conditions; e.g., liquid-type reactors include suspended-growth, fixed-film, and submerged fixed-film reactors. Solids can be treated using a slurry phase-type reactor.

Bioremediation Bioremedition is the process by which organic or inorganic waste is biologically degraded or transformed, usually to innocuous materials; the process can occur naturally or can be enhanced by adding an electron acceptor, nutrients, or other factors which would otherwise limit the natural biodegradation rate.

Bioventing Bioventing is the process by which contaminants in the unsaturated zone are removed by volatilization and biodegradation as air or oxygen is supplied by vacuum extraction and/or injection.

Bodenfilter A bodenfilter is a biofilter consisting of one or more beds of biologically active material, primarily mixtures based on compost, peat or soil through which gas is vented. Aerobic degradation of contaminants in the gas occurs in the biofilm if microorganisms are present that can metabolize them.

Cometabolism Cometabolism is the process by which an organic compound is metabolized or transformed but not used for growth.

Composting Composting is a procedure for accelerating biodegradation of contaminants at elevated temperatures by aerating and adding bulking agents and possibly nutrients to waste in a compost pile.

Congener A chemical compound closely related to another in molecular structure is a congener; e.g, PCB molecules may differ from each other by the location or number of chlorine atoms attached to the carbon ring structure.

Constitutive mutant or strain A constitutive mutant or strain is a population of organisms in which the enzyme controlling a process is functional without having to be induced (see enzyme).

Darcy's Law Darcy's Law is expressed by an equation that can be used to compute the quantity of water flowing through an aquifer; hydraulic conductivity is a parameter in the equation.

Dehalogenation Dehalogenation is a chemical reaction in which a halogen atom is replaced by a proton; reductive dehalogenation (dechlorination) of trichloroethylene results in the removal of one chlorine and leaves as the other product dichloroethylene.

Denitrification Denitrification is a microbial process that uses nitrate as a terminal electron acceptor to oxidize an organic compound.

Enzyme An enzyme is a proteinaceous biochemical catalyst; a biological reaction is controlled and accelerated by an enzyme; most enzymes are substrate-specific with a characteristic affinity for attaching to the substrate to mediate the reaction. Induction of an enzyme is a process by which synthesis of enzymes required to metabolize a substrate is initiated by the inactivation of a repressor substance.

Ex situ Ex situ refers to the execution of an environmental cleanup by removing the contaminants from the existing location to another matrix, either on- or off site for treatment, e.g., bioreactors.

Geochemistry Geochemistry is the science of the chemical composition of the earth and the processes that cause the distribution of the elements.

Groundwater Groundwater is subsurface water, especially that water found in the saturated zone below the water table, and is found in formations known as aquifers.

Heterotrophic organisms Heterotrophic organisms derive their energy and carbon required for growth from organic compounds, as opposed to autotrophic organisms which can grow and obtain energy from inorganic nutrients.

Hydrocarbon A hydrocarbon is a chemical that is composed of only carbon and hydrogen, e.g., gasoline, benzene, methane.

Hydraulic conductivity Hydraulic conductivity is a coefficient of proportionality describing the rate at which water can move through a permeable medium (See Darcy's Law).

Inorganic substance An inorganic substance is a chemical that does not contain carbon-to-hydrogen bonds, e.g. metals, nitrate, phosphate.

In situ In situ means "in place" in the environment.

Land treatment Land treatment of wastes and environmental contaminants is a general term for processes that accelerate biodegradation of contaminants in situ or ex situ by aerating and possibly adding nutrients to contaminated surface soil or contaminated material that has been applied to soil.

Metabolism Metabolism is the chemical reaction in an organism resulting in the breakdown of substances to produce energy and growth; products of the metabolic process are called metabolites; catabolic metabolism is a degradative process.

Methanogen Methanogens are strict anaerobic bacteria that produce methane while growing on CO_2-type substrates (CO_2, HCOOH, CO), methyl-containing compounds (CH_3OH, $CH_3NH_3^+$, $(CH_3)_2NH_2^+$, $(CH_3)_3NH^+$) or acetic acid.

Methanotroph (methylotrophs) Methanotrophs are aerobic bacteria that use methane and other one- and two-carbon compounds as a carbon source; these organisms may also be capable of cometabolic processes that degrade contaminants, such as chlorinated ethenes.

Michaelis-Menten kinetics Michaelis-Menten kinetics describe reaction rate dynamics of enzyme-controlled chemical processes; the equation expresses the relationship between reaction rate and substrate concentration and is dependent on the relative affinity of an enzyme for a substrate.

Mineralization Mineralization is the process of decomposition of an organic compound to inorganic products.

Monod kinetics Monod kinetics is the application of the Michaelis-Menten-type equation to describe the growth dynamics of a culture of microorganisms, where growth rate is related to substrate concentration.

Mutant strain A mutant strain is a population of organisms genetically different from their ancestors due to chromasomal changes.

Organic compound An organic compound contains carbon; the exception, carbon dioxide, is generally considered inorganic.

Oxidation, oxidant In a chemical reaction, oxidation is the the loss of an electron by a reactant.

Oxygenase enzyme An oxygenase enzyme incorporates one or both atoms of molecular oxygen (O_2) into an organic compound to yield hydroxyl groups.

Permeability Permeability, or intrinsic permeability, is a property of a porous medium independent of the nature of the liquid and pertains to the relative ease with which a porous medium can transmit a liquid under a hydraulic or potential gradient.

Plating technique Plating is a microbiological technique to enumerate microorganisms which results in a "viable count;" the microorganisms are uniformly distributed in or on a growth medium which commonly contains agar in shallow containers called Petri dishes; the inoculated microorganisms grow into colonies visible without magnification.

Porosity Porosity is a property of porous medium; it is the ratio of the volume of void spaces in a rock or sediment to the total volume.

Redox Redox is a chemical process, an oxidation-reduction reaction; the loss (oxidation) and gain (reduction) of electrons among reactants affect the charge of the medium and is expressed as electrical potential, Eh (volts) for a particular hydrogen ion concentration, pH; most oxidizing environments have a Eh level greater than +400 mV.

Reduction, reductant In a chemical reaction, reduction is the gain of an electron by a substance. A reductant is a substance that induces the gain of an electron by another substance.

Remedial Investigation (RI) The remedial investigation is part of the Superfund remediation process of a hazardous waste site in which the site conditions are characterized and appropriate treatability studies are performed to evaluate potential cleanup technologies; risk assessment of exposure and toxicity assessment are initiated during this time.

Reynold's number The Reynold's number is a dimensionless number used to determine whether flow will be turbulent or laminar; in laminar flow the particles follow paths that are smooth, straight, and parallel to the channel walls.

Saturated zone The saturated zone is the subsurface region, an aquifer, in which the voids in the rock or solids are filled with water at a pressure greater than atmospheric. The water table is the top of the saturated zone in an unconfined aquifer.

Seepage velocity The seepage velocity, also called average velocity, is the actual rate of movement of fluid particles through porous media. The velocity, taking into account the effective porosity of the medium, is calculated using Darcy's Law.

Soil Vapor Extraction (SVE) Test Soil vapor extraction test is a site characterization test to pull vapor from the unsaturated zone of the subsurface to analyze for volatile contaminants to evaluate the type and extent of contamination.

Soil venting Soil venting is a type of remedial technology applied to the unsaturated zone of the subsurface in which vapor is forced out of the formation, or displaced with air and the volatile organic contaminants are removed from the effluent by sorption.

Stoichiometry Stoichiometry is the method of calculation of the combining amounts of reactants and products in a chemical reaction.

Surfactant Surfactant is a term formed from the contraction of "surface-active agent;" surfactants are ionic or nonionic organic compounds with amphipatic structure, i.e, the molecule is composed of groups of opposing solubility tendencies, typically an oil-soluble (hydrophobic) and a water-soluble, ionic or polar group (hydrophilic) section.

Transmissivity Transmissivity refers to the rate at which water is transmitted through an aquifer under a particular hydraulic gradient due to the properties of the porous media and its thickness.

Unsaturated zone The unsaturated (vadose) zone is located between the land surface and the water table. It includes the root zone (rhizosphere), intermediate zone, and capillary fringe of an aquifer. The unsaturated zone pore spaces contain water at less than atmospheric pressure, as well as air and other gases.

Xenobiotic compound A xenobiotic compound is a synthetic compound.

C

LIST OF ACRONYMS

BTEX	Benzene, toluene, ethylxylene, xylenes
DNAPL	Dense nonaqueous phase liquid
EPA	Environmental Protection Agency
GC-FID/PID	Gas chromatography - flame ionization detector/photoionization detector
GC-FID/MS	Gas chromatography - flame ionization detector/mass spectrophotometer
HDPE	High density polyethylene
LNAPL	Less dense nonaqueous phase liquid
MTBE	Methyl tertiary butyl ether
NAPL	Nonaqueous phase liquid
PAH	Polynuclear or polycyclic aromatic hydrocarbon
PCE	Perchloroethylene (Tetracholorethylene)
PCP	Pentachlorophenol
PVC	Polyvinylchloride
RCRA	Resource Conservation and Recovery Act 1976
RI	Remedial investigation
SVE	Soil vapor extraction
TCE	Trichloroethylene
TKN	Total Kjeldahl nitrogen
TNT	trinitrotoluene
TOC	Total organic carbon

TP	Total phosphorus
TPH	Total petroleum hydrocarbon
VES	Vapor extraction system
ZOI	Zone of Incorporation

D

LIST OF REFERENCES

Acton, D.W. and J.F. Barker. 1992. In situ biodegradation potential of aromatic hydrocarbons in anaerobic groundwater. *J. Contam. Hydrol.* 9:325.

Aggrawal, P.K., J.L. Means and R.E. Hinchee. 1992. Formulation of nutrient solutions for in situ bioremediation. In *In situ bioreclamation: applications and investigations for hydrocarbon and contaminated site remediation*, 51. Ed. R.E. Hinchee and R.F. Olfenbuttel. Stoneham, Mass: Butterworth-Heinemann.

Alexander, M. 1994. *Biodegradation and Bioremediation.* San Diego: Academic Press.

American Petroleum Institute. 1983. *Land treatment practices in the petroleum industry.* API: Washington, D.C.

Apajalahti, J.H.A. and M.S. Salkinoja-Salonen. 1986. Dechlorination and *para*-hydroxylaltion of polychlorinated phenols by *rhodococcus chlorophenolicus. J. Bacteriol.* 169:675.

Appendix C: Biological treatment technologies. 1988. *Technology screening guide for treatment of CERCLA soils and sludges,* 103. EPA/540/2-88/004. September.

Arciero, D., T. Vannelli, N. Logan, and A.B. Hooper 1989. Degradation of trichloroethylene by the ammonia oxiding bacterium *Nitrosomonas europaea. Biochem. Biophys. Res. Commun.* 159:6.

Bak, F. and F. Widdel. 1992. In situ biodegradation of phenol and phenol derivatives by *Desulfobacterium phenolicum* sp. nov. *Arch. Microbiol.* 146:177.

Barcelona, M.J. and T.R. Holm. 1991. Oxidation-reduction capacities of aquifer solids. *Env. Sci. Technol.* 25:1565.

Barker, J.F., G.C. Patrick, and D. Major. 1987. Natural attenuation of aromatic hydrocarbons in a shallow sand aquifer. *Ground Water Mon. Rev.* Winter:64.

Bartha, R. and I. Bossert. 1984. Treatment and Disposal of Petroleum Refinery Wastes. In *Petroleum microbiology*, 553. Ed. R.M. Atlas. New York: MacMillan Publishing Co.

Bartha, R. and J.T. Dibble. 1979. Effect of environmental parameters on the biodegradation of oil sludge. *Appl. Environ. Microbiol.* 37:729.

Battermann, G. 1986. Decontamination of polluted aquifers by biodegradation. In *1985 Intl. TNO Conf. on Contaminated Soil,* 711. Ed. J.W. Assink and W.J. van den Brink. Nijhoff, Dordrecht.

Bedard, D.L., M.L. Haberl. R.J. May., and M.J. Brennan. 1987. Evidence for novel mechanisms of polychlorinated biphenyl metabolism in *Alcaligenes eutrophus* H850. *Appl. Envi. Microbiol.* 53:1103.

Bedard, D.L. and J.F. Quensenl (in press). Microbial reductive dechlorination of polychlorinated biphenyls. In: *Microbial Transformation and Degradation of Toxic Organic Chemcials, 121-209.* L.Y Yound and C. Cerniglia, eds. John Wiley & Sons.

Bell R.A. and A.H. Hoffman. 1991. Gasoline spill in fractured bedrock addressed with in situ bioremediation, 437-443. In *In Situ Bioremediation,* R.E. Hinchee and R.F. Olfenbuttle, eds. Butterworth-Heinemann, Stoneham, Mass.

Beller, H.R., D. Grbic-Galic, and M. Reinhard. 1992. Microbial degradation of toluene under sulfate reducing conditions and the influence of iron on the process. *Appl. and Environ. Microbiol.* 58:786.

Bennedsen, M.G., J.P. Scott, and J.D. Hartley. 1987. Use of vapor extraction systems for in situ removal of volatile organic compounds from soil. In *Proc. of Natl. Conf. on Hazardous Wastes and Hazardous Materials,* 92.. Washington, D.C.

Bohn, H. and R. Bohn. 1988. Soil beds weed out air pollutants. *Chem. Engr.* Apr. 25:73-6.

Bonin, H., M. Guillerme, P. Lecomte, and M. Manfredi. 1994. Optimizing in situ bioremediation of a pilot site. In *Hydrocarbon Bioremediation*, 252. Eds. R.E. Hinchee, B.C. Alleman, R.E. Hoeppel, and R.N. Miller. Boca Raton: Lewis Publishers.

Borden, R.C. and P.B. Bedient. 1986. Transport of dissolved hydrocarbons influenced by oxygen-limited biodegradation, 1. theoretical development. *Water Res. Res.* 22:1973.

Borden, R.C., M.D. Lee, J.M. Thomas, P.B.Bedient, and C.H. Ward. 1989. In situ measurement and numerical simulation of oxygen limited biotransformation. *Ground Water Mon. Rev.* 9:83.

Borden, R.C., P.B. Bedient, M.D. Lee, C.H. Ward, and J.T. Wilson. 1986. Transport of dissolved hydrocarbons influenced by oxygen-limited biodegradation 2. field application. *Water Res. Res.* 22:1983.

Bossert, I.D. and L.Y. Young. 1986. Anaerobic oxidation of *p*-cresol by a denitrifying bacterium. *Appl. Environ. Microbiol.* 52:1117.

Bourquin, A.W. 1989. Bioremediation of hazardous waste. *Haz. Mat. Con.* 2:16.

Bouwer, E.J. 1992. Bioremediation of organic contaminants in the subsurface. In *Environmental microbiology,* 287. Ed. Ralph Mitchell. New York: Wiley-Liss.

Bouwer, H. 1984. Elements of soil science and groundwater hydrology. In *Groundwater pollution microbiology*, 9. Ed. G. Bitton and C.P. Gerba. New York: John Wiley & Sons.

Boyd, S.A. and D.R. Shelton. 1984. Anaerobic biodegradation of chlorophenols in fresh and acclimated sludge. *Appl. Environ. Microbiol.* 47:272.

Brown, J.F., D.L. Bedard, M.J. Brennan, J.C. Carnahan, H. Feng, and R.E. Wagner. 1987. Polychlorinated biphenyl dechlorination in aquatic sediments. *Science* 241:709.

Brown, K.W., G.B. Evans, and B.D. Frentrup. 1983. *Hazardous waste land treatment.* Woburn, MA: Butterworth Publishers.

Brown, R.A. and R.D. Norris. 1994. The evolution of a technology: hydrogen peroxide in in situ bioremediation. In *Hydrocarbon Bioremediation*, 148. Eds. R.E. Hinchee, B.C. Alleman, R.E. Hoeppel, and R.N. Miller. Boca Raton, FL: Lewis Publishers.

Brown, R.A. 1993. Air sparging. In *In situ bioremediation of groundwater and geological material: a review of technologies.* Office of Research and Development, U.S. Environmental Protection Agency, R.S Kerr Environmental Research Laboratory: Ada, Okla.

Brown, R.A. and F. Jasinlewicz. 1992. Air sparging: a new model for remediation. *Poll. Eng.* 24:52.

Brown, R.A. and R. Fraxedas. 1992. Air sparging—extending volatilization to contaminated aquifers. In *Proc. Symp. on Soil Venting*, 249. U.S. Environmental Protection Agency, R.S. Kerr Environmental Research Laboratory, National Center for Ground Water Research. April 29-May, 1991. Houston, Texas.

Brox, G.H. and D.E. Hanify. 1989. A new solid/liquid contact bioslurry reactor making bio-remediation more cost-competitive. Paper presented at *Colorado Hazardous Waste Management Society Conf.* Denver, Colo.

Brubaker, G.R. and H.F. Stroo. 1992. In situ bioremediation of aquifers containing polyaromatic hydrocarbons. *J. Haz. Mater.* 32:163.

Brunner, W., D. Staub, and T. Leisinger. 1980. Bacterial degradation of dichloromethane. *Appl. Environ. Microbiol.* 40:950.

Bulger, P.R., A.E. Kehew, and R.A. Nelson. 1989. Dissimilatory nitrate reduction in a waste-water contaminated aquifer. *Ground Water* 27:664.

Carlson, D.A. and C.P. Leiser. 1966. Soil beds for the control of sewage odors. *J. Wat. Poll. Contr. Fed.* 38:829.

Chiang, C.Y., J.P. Salanitro, E.Y. Chai, J.D. Colthart, and C.L. Klein. 1989. Aerobic biodegradation of benzene, toluene, and xylene in a sandy aquifer-data analysis and computer modeling. *Ground Water* 27:823.

Cho, J.S. and D.C. DiGiulio. 1992. Pneumatic pumping testing for soil vacuum extraction. *Environ. Prog.* 11:228.

Compeau, G.C., W.D. Mahaffey, and L. Patras. 1991. Full-scale bioremediation of contamination soil and water. In *Environmental biotechnology for waste treatment.* New York: Plenum Press.

Cozzarelli, I. M., R.P. Eganhouse, and M.J. Baedecker. 1990. Transformation of monoaromatic hydrocarbons to organic acids in anoxic groundwater environment. *Environ. Geol. Water Science.* 16:135.

Crawford, R.L. and W. Mohn. 1985. Microbiological removal of pentachlorophenol from a soil using a *Flavobacterium. Enzyme Microb. Technol.* 7:617.

Cunningham, J. 1992. Telephone conversation with M.R. Piotrowski. 13 August.

Davis-Hoover, W.J., L.C. Murdoch, S.J. Vesper, H.R. Pahren, O.L. Sprockel, C.L. Chang, A. Hussain, and W.A Ritschel. 1991. Hydraulic fracturing to improve nutrient and oxygen delivery in in situ bioreclamation. In *In Situ Bioreclamation: Applications and Investigations for Hydrocarbon and Contaminated Site Remediation*, 67. Eds. R.E. Hinchee and R.F. Olfenbuttel. Stoneham, MA: Butterworth-Heinemann.

deBont, J.A.M., J.A. Vorage, S. Hartmans, and W.J.J. van den Tweel. 1986. Microbial degradation of 1,3-dichlorobenzene. *Appl. Environ. Microbiol.* 52:677.

deBruin, W.P., M.J.J. Kotterman, M.A. Posthumus, G. Schraa, and A.J.B. Zehnder. 1992. Complete bilogical reductive transformation of tetrachloroethane to ethane. *Appl. Environ. Microbiol.* 58:1996.

Dickel, O. and H-J. Knackmuss. 1991. Catabolism of 1,3-dinitrobenzene by *Rhodococcus* sp. QT-1. *Arch. Microbiol.* 157:76.

DiStefano, T.D., J.M. Gossett, and S.H. Zinder. 1991. Reductive dechlorination of high concentrations of tetrachloroethane to ethane by an anaerobic enrichment culture in the absence of methanogenesis. *Appl. Environ. Microbiol.* 57:2287.

DiStefano, T.D., J.M. Gossett, and S.H. Zinder. 1992. Hydrogen as an electron donor for dechlorination of tetrachloroethene by an anaerobic mixed culture. *Appl. Environ. Microbiol.* 58:3622.

Downey, D.C. and P.R. Guest, 1991. Physical and biological treatment of deep diesel-contaminated soils. In Proc. *Petroleum hydrocarbons and organic chemicals in groundwater: prevention, detection and restoration,* 361. National Well Water Assocation/American Petroleum Institute. Houston, Texas.

Downey, D.C., J.F. Hall, and R.N. Miller. 1992. Bioventing in low permeability soils. In *Proc. 6th Natl. Outdoor Action Conf,* 599. May 11-13. Las Vegas. Dublin, OH: National Ground Water Association.

DuPont, R.R. 1992a. Application of bioremediation fundamentals to the design and evaluation of in situ soil bioventing systems. Paper presented at *85th Annual Meeting & Exhibition of the Air & Waste Management Association.* Paper 92-30.03. Kansas City, Mo.

DuPont, R.R. 1992b. Nitrogen fixation potential in petroleum contaminated soils. Paper presented at *85th Annual Meeting & Exhibition of the Air & Waste Management Association.* Paper 92.13.06. Kansas City, Mo.

DuPont, R.R., W. Doucette and R.E. Hinchee. 1991. Assessment of in situ bioremediation potential and the application of bioventing at a fuel-contaminated site. In *In situ bioreclamation, applications and investigations for hydrocarbon and contaminated site remediation,* 262. Ed. R.E. Hinchee and R.F. Olfenbuttel. Stoneham, Mass: Butterworth-Heinemann.

Eckenfelder, W.W., Jr. 1967. *Industrial water pollution control.* New York: McGraw-Hill Book Co.

Edwards, E.A. and D. Grbic-Galic. 1992. Complete mineralization of benzene by aquifer microorganisms under strictly anaerobic conditions. *Appl. Environ. Microbiol.* 58:2663.

Egli, C., R. Scholtz, A.M. Cook, and T. Leisinger. 1987. Anaerobic dechlorination of tetrachloroethane and 1,2-dichloroethane to degradable products by pure cultures of *Desulfobacterium* sp.and *Methanobacterium* sp. *FEMS Microbiol. Lett.* 43:257.

Ehlke, T.A., T.E. Imbrigiotta, B.H. Wilson, and J.T. Wilson. 1991. Biotransformation of cis-1,2-dichloroethylene in aquifer material from Picatinny Arsenal, Morris County, New Jersy. In *Proc. Technical Meeting: U.S. Geological Survey Toxic Substances Hydrology Program,* 689. U.S. Geological Survey Water Resources Investigations Report 91-4034. Monterey, Calif.

Ely, E.L. and D.A. Heffner. 1988. *Process of in situ biodegradation of hydrocarbon contaminated soil.* U.S. Patent Number 4,765,902.

Ensley, B.D. 1991. Biochemical diversity of trichloroethylene metabolism. *Ann. Rev. Microbiol.* 45:283.

Environmental Protection Agency. 1991. Vendor Information system for Innovative Treatment Technologies (VISITT). Office of Solid Waste and Emergency Response, Washington, DC, EPA/540/2-91/001, June 1991.

Ewers, J., D. Freire-Schroder, and H.J. Knackmuss. 1990. Selection of (TCE) trichloroethene degrading bacteria that resist inactivation by TCE. *Arch. Microbiol.* 154:410.

Fathepure, B.Z. and T.M. Vogel. 1991. Complete degradation of polychlorinated hydrocarbons by a two-stage biofilm reactor. *Appl. Environ. Microbiol.* 57:3418.

Fathepure, B.Z., and S.A. Boyd. 1988. Dependence of tetrachloroethylene dechlorination on methanogenic substrate consumption by *Methanosarcina* spl strain DCM. *Appl. Environ. Microbiol.* 54:2,976.

Federal Register. 1984. 49:40. (October 26).

Fernando, T., J.A. Bumpus, and S.D. Aust. 1990. Biodegradation of TNT (2,4,6-trinitrotoluene) by *Phanerochaete chrysosporium. Appl. Environ. Microbiol.* 56:166.

Fiorenza, S. 1991. Use of hydrogen peroxide for subsurface remediation: microbial responses and their implications. Rice University thesis, 296 pp.

Fortnagel, P., H. Harms, R.M. Wittich, S. Krohn, H. Meyer, V. Sinnwell, H. Wilkes, and W. Francke. 1990. Metabolism of dibenzofuran by *Pseudomonas* sp. strain HH69 and the mixed culture HH27. *Appl. Environ. Microbiol.* 56:1148.

Fox, B.G., J.G. Borneman, L.P. Wackett, and J.D. Lipscomb. 1990. Haloalkene oxidation by the soluble methane monooxygenase from *Methylosinus trichosporium* OB3b: mechanistic and environmental applications. *Biochem.* 29:6419.

Francy, D.S., J.M. Thomas, R.L. Raymond, and C.H. Ward. 1991. Emulsification of hydrocarbons by subsurface bacteria. *J. Ind. Microbiol.* 8:237.

Freedman, D.L. and J.M. Gossett. 1989. Biological reductive dechlorination of tetrachloroethylene and trichloroethylene to ethylene under methanogenic conditions. *Appl. Environ. Microbiol.* 55:2144.

Freedman, D.L. and J.M. Gossett. 1991. Biodegradation of dichloromethane and its utilization as a growth substrate under methanogenic conditions. *Appl. Environ. Microbiol.* 57:2847.

Gibson, D.T. 1984. *Microbial degradation of organic compounds.* New York: Marcel Dekker, Inc.

Godsy, E.M. 1987. Conversation with M.R. Piotrowski. June 11.

Godsy, E.M., D.F. Goerlitz, and D. Grbic-Galic. 1992a. Methanogenic biodegradation of creosote contamination in natural and simulated ground-water ecosystems. *Ground Water* 30:232.

Godsy, E.M., D.F. Goerlitz, and D. Grbic-Galic. 1992b. Methanogenic degradation kinetics of phenolic compounds in aquifer-derived microcosms. *Biodegradation* (in press).

Godsy, E.M., D.F. Goerlitz, and G.G. Ehrlich. 1983. Methanogenesis of phenolic compounds by a bacterial consortium from a contaminated aquifer in St. Louis Park, Minnesota. *Bull. Environ. Contam. Toxicol.* 30:261.

Golueke, C.G. and L.F. Diaz. 1989. Biological treatment for hazardous wastes. *BioCycle* 30:58.

Grbic-Galic, D. 1990. Anaerobic microbial transformation of nonoxygenated aromatic and alicyclic compounds in soil, subsurface and freshwater sediments. In *Soil Biochemistry,* 117. Ed. J.M. Bollag and G. Stotzky. New York: Marcel Dekker, Inc.

Hadley, P.W. and R. Armstrong. 1991. Where's the benzene?-examining California ground-water quality surveys. *Ground Water* 29:35.

Haigler, B.E. and J.C. Spain. 1989. Degradation of 1,2-dichlorobenzene by a *Pseudomonas* sp. *Appl. Environ. Microbiol.* 54:294.

Hartmans, S. and J.A.M. deBont. 1992. Aerobic vinyl chloride metabolism in *Mycobacterium aurum* L1. *Appl. Environ. Microbiol.* 58:1220.

Hazen, T.C. 1991. *Test plan for in situ bioremediation demonstration of the Savannah River integrated demonstration project.* DOE/OTD TTP NO.SR 0566-01, WSRC-RD-91-23.

Hearn, P.P., Z.A. Brown, and K.O. Denning. 1986. Analysis of sand grain coatings and major-oxide composition of samples from a creosote works, Pensacola, Florida. In *Movement and Fate of Creosote Wastes in Ground Water, Pensacola, Florida.* U.S. Geological Survey Water Supply Paper 2285. Alexandria, VA: U.S. Geological Survey.

Hildebrandt, W.W. and S.B. Wilson. 1990. On-site remediation of organic impacted soils on oil field properties. Paper presented at *California Regional Meeting of the Society of Petroleum Engineers.* Ventura, Calif. April.

Hinchee, R.E. 1993. Bioventing of petroleum hydrocarbons. In *In situ bioremediation of groundwater and geological materials: a review of technologies.* Office of Research and Devopment., U.S. Environmental Protection Agency, R.S. Kerr Environvironmental Research Laboratory. Ada, Okla.

Hinchee, R.E. and S.K. Ong. 1992. A rapid in situ respiration test for measuring aerobic biodegradation rates of hydrocarbons in soil. *J. Air Waste Manage. Assoc.* 42:1309.

Hinchee, R.E., S.K. Ong, R.N. Miller, D.C. Downey, R. Frandt. 1992. *Test plan and technical protocol for a field treatability test for bioventing*. (PB93-209146). Air Force Center for Environmental Excellence. Brooks Air Force Base, Texas.

Hinchee, R.E., D.C. Downey, R.R. DuPont, P. Aggarwal and R.N. Miller. 1991. Enhancing biodegradation of petroleum hydrocarbon through soil venting. *J. Haz. Mat.* 27:315.

Hogg, D.S., R.J. Burden, P.J. Riddell. 1992. In situ vadose zone bioremediaiton of soil contaminated with nonvolatile hydrocarbons. Presented at *HMCRI Conference.* San Francisco. Feb.4.

Hopper, D.J. 1976. The hydroxylation of *p*-cresol and its conversion to *p*-hydroxybenzaldehyde in *Pseudonomas putida. Biochem. Biophys. Res. Commun.* 48:462-8.

Hopper, D.J. 1978. Incorporation of [^{18}O] water in the formation of *p*-hydroxybenzyl alcohol by the *p*-cresol methylhydroxylase from *Pseudonomas putida. J. Biochem.* 175:345-7.

Howard, P.H. 1989. *Handbook of environmental fate and exposure data for organic chemicals: volume I large production and priority pollutants*. Chelsea, Mich: Lewis Publishers.

Howard, P.H. 1990. *Handbook of environmental fate and exposure data for organic chemicals: volume II solvents*. Chelsea, Mich: Lewis Publishers.

Howard, P.H., R.S. Boethling, W.F. Jarvis, W.M. Meylan, and E.M. Michalenko. 1991. *Handbook of environmental degradation rates*. Chelsea, Mich: Lewis Publishers.

Hurlburt, S. 1987. Raymond approaches hydrocarbon spills head-on. *Ground Water Mon. Rev*. Spring:90.

Hutchins, S.R. and J.T. Wilson. 1991. Laboratory and field studies on BTEX biodegradation in a fuel-contaminated aquifer under denitrifying conditions. In *In situ bioreclamation: applications and investigations for hydrocarbon and contaminated site remediation,* 157. Ed. R.E.Hinchee and R.F. Olfenbuttel. Stoneham, Mass: Butterworth-Heinemann.

Hutchins, S.R., G.W. Sewell, D.A. Kovacs, and G.A. Smith. 1991b. Biodegradation of aromatic hydrocarbons by aquifer microorganisms under denitrifying conditions. *Environ. Sci. Technol.* 25:68.

Hutchins, S.R., W.C. Downs, J.T. Wilson, G.B. Smith, D.A. Kovacs, D.D. Fine, R.H. Douglas, and D.J. Hendrix. 1991a. Effect of nitrate addition on biorestoration of fuel-contaminated aquifer: field demonstration. *Ground Water* 29:571.

Hutzler, N.J., Baillod, C.R., and P.A. Schaepe. 1989. Biological reclamation of soils contaminated with pentachlorophenol, 361. In Proc. *Sixth National Conference on Hazardous Wastes.* Louisiana. April.

Imbrigotta, T.E., T.A. Ehlke, and M. Martin. 1991. Chemical evidence of processes affecting the fate and transport of chlorinated solvents in ground water at Picatinny Arsenal, New Jersey, 681-8. In Proc. *Technical Meeting: U.S. Geological Survey Toxic Substances Hydrology Program.* U.S. Geological Survey Water Resources Investigations Report 91-4034. Monterey, Calif. March 11-15.

Johnson, P.C. and R.A. Ettinger. 1992. "Some considerations for the design of in situ vapor extraction systems: radius of influence-vs-radius of remediation. Paper presented at *3rd Intl. Conf. on Ground Water Quality Research.* Dallas. June 21-24.

Kampbell, D.H., J.T. Wilson, H.W. Read, and T.T. Stocksdale. 1987. Removal of volatile aliphatic hydrocarbons in a soil bioreactor. *J. Air Poll. Con. Assn.* 37:1236.

Keeney, D. 1986. Sources of nitrate to groundwater. *CRC Crit. Rev. Environ. Con.* 16:257.

Kemp, P.F., B.F. Sherr, E.B. Sherr, and J.J. Cole. 1993. *Handbook of Methods in Aquatic Microbial Ecology.* Boca Raton, FL: Lewis Publishers.

Klecka, G.M., J.W. Davis, D.R. Gray, and S.S. Madsen. 1990. Natural bioremediation of organic contaminants in groundwater: Cliffs-Dow Superfund Site. *Ground Water* 28:534.

Kuhn, E.P., G.T. Townsend, and J.M. Suflita. 1990. Effect of sulfate and organic carbon supplements on reductive dehalogenation of chloroanilines anaerobic aquifer slurries. *Appl. Environ. Microb.* 56:2630.

Kuhn, E.P., J. Zeyer, P.Eicher, and R.P. Schwarzenbach. 1988. Anaerobic degradation of alkylated benzenes in denitrifying laboratory aquifer columns. *Appl. Environ. Microbiol.* 54:490-6.

Leavitt, M.E., C.A. Lang, J. Sanseverino, K. Hague, D. Dameron, J. Rightmyer, and D.A. Graves. 1992. Implications of surfactant augmentation for in situ bioremediation systems. In *Proceedings from the Air and Waste Management Association*, 75. Pittsburgh, PA: Air and Waste Management Association.

Leavitt, M.E. and J.F. Jadlocki. 1989. In Situ Bioremediation of Diesel Fuel in Unsaturated Soils: Results of a Laboratory Study. In: *Hazardous waste treatment: biosystems for pollution control*, 283. Air and Waste Management Association. Pittsburgh.

Lee, M.D., J.T. Wilson, and C.H. Ward. 1987. In situ restoration techniques for aquifers contaminated with hazardous wastes. *J. Hazard. Mater.* 14:71.

Lee, M.D., J.M. Thomas, R.C. Borden, P.B. Bedient, J.T. Wilson, and C.H. Ward. 1989. Biorestoration of aquifers contaminated with organic compounds. *CRC Crit. Rev. Env. Contr.* 18: 29.

Leeson, A., R.E. Hinchee, J. Kittel, G. Sayles, C.M Voges, and R.N. Miller. 1994. Optimizing bioventing in shallow vadose zones and cold climates. *Hydrol. Sci.* 38:283-295.

Lemon, L.A., J.R. Barbaro, and J.F. Barker. 1988. Biotransformation of BTEX under anaerobic denitrifying conditions: evaluation of of field observations. In *Proc. FOCUS Conf. on Eastern Regional Ground Water Issues,* 213.. National Water Well Association. Dublin, Ohio.

Lenke, H., P.G. Rieger and H.J. Knackmuss. 1992. Reductive mechanism in the aerobic bacterial degradation of picric acid. In *Abst. Ann. Mtg. Amer. Soc. Microbiol.* Abstract No. 0148.

Leson, G. and A.M. Winer. 1991. Biofiltration: an innovative air pollution control technology for VOC emissions. *J. Air Waste Man. Assoc.* 41:1045.

Li, S. and L.P.Wackett. 1992. Trichloroethylene oxidation by toluene dioxygenase. *Biochem. Biophys. Res. Comm.* (in press).

Linkenheil, R.J. and T.J. Patnode. 1987. Bioremediation of contamination by heavy organics at a wood preserving plant site, 193. In Proc. *Superfund '87 8th National Conference.* Washington D.C. Nov.

Loehr, R.C. and J.F. Malina, Jr. 1986. Land treatment: a hazardous waste management alternative. Paper presented at *Water Resources Symp. #13.* Center for Research in Water Resources. The University of Texas at Austin. Austin.

Loehr, R.C., D.C. Erickson, L.A. Rogers, and D.M. Kelmar. 1990. *Mobility and degradation of residues at hazardous waste land treatment sites at closure.* EPA/600/2-90/018. Robert S. Kerr Environmental Ressearch Laboratory, Office of Research Development, U.S. Environmental Protection Agency. Ada, Okla.

Loehr, R.C., J.H. Martin, and E.F. Neuhauser. 1992. Land treatment of an aged oily sludge-organic loss and change in soil characteristics. *Water Res.* 26:805.

Lovley, D.R. 1991. Dissimilatory Fe(III) and Mn(IV) reduction. *Microbiol. Rev.* 55:259.

Lovley, D.R. and D.J. Lonergan. 1990. Anaerobic oxidation of toluene, phenol, and *p*-cresol by the dissimilatory iron-reducing organism, GS-15. *Appl. Environ. Microbiol.* 56:1858.

Lovley, D.R., and E.J.P. Phillips. 1988. Novel mode of microbial energy metabolism; organic carbon oxidation coupled to dissimilatory reduction of iron or manganese. *Appl. Environ. Microbiol.* 54:1472.

Lowes, B.C. 1991. Soil-induced decomposition of hydrogen peroxide. In *In situ bioreclamation: application and investigation for hydrocarbons and contaminated site remediation,* 143. Ed. R.E. Hinchee and R.F. Olfenbuttel. Stoneham, Mass: Butterworth-Heinemann.

Lynch, J. and B.R. Genes. 1988. Land treatment of hydrocarbon contaminated soils. In *Petroleum Contaminated Soils,* 175. Vol. 1. Ed. P.T. Kostecki and E.J. Calabrese. Chelsea, Mich:Lewis Publishers, Inc.

Marley, M.C. and G.E. Hoag. 1984. Induced soil venting for recovery/restoration of gasoline hydrocarbons in the vadose zone. In *Proc. Petroleum Hydrocarbons and Organic Chemicals in Groundwater,* 473. National Water Well Association/American Petroleum Institute. Houston.

McCarthy, J.F. 1989. Bioavailability and toxicity of metals and hydrophobic organic contaminants. In *Aquatic humic substances: influence on fate and treatment of pollutants,* 263. Ed. I.H. Suffet and P. MacCarthy. American Chemical Society. Washington, D.C.

McCarty, P.M., L. Semprini, and P. Roberts. 1989. In-situ biotransformation methodologies. In *In-Situ Aquifer Restoration of Chlorinated Aliphatics by Methanotrophic Bacteria*, 197. Stanford University Department of Civil Engineering Technical Report No. 310.

McCormick, N.G., F.F. Feeherry, and H.S. Levinson. 1976. Microbial transformation of 2,4,6-trinitrotoluene and other nitroaromatic compounds. *Appl. Environ. Microbiol.* 31:949.

McCormick, N.G., J.H. Cornell, and A.M. Kaplan. 1978. Identification of biotransformation products from 2,4-dinitrotoluene. *Appl. Environ. Microbiol.* 35:945

McKenna, E.J. and R.D. Heath. 1976. *Biodegradation of polynuclear aromatic hydrocarbon pollution by soil and water microorganisms.* UILU-WRC-76-0113. University of Illinois Research Report.

Michaelson, D.L. and M. Lofti. 1990. Oxygen microbubbles injection for in situ bioremediation: possible field scenarios. In *Innovative Hazardous Waste Systems*. New York, NY: Technotric.

Mihelcic, J.R. and R.G. Luthy. 1988. Microbial degradation of acenaphthene and naphthalene under denitrification conditions in soil-water systems. *Appl. Environ. Microbiol.* 54:1188.

Mihelcic, J.R., D.R. Lueking, R.J. Mitzell, and J.M. Stapleton. 1993. Bioavailability of sorbed- and separate-phase chemicals. Biodegradation 4:141-53.

Mikesell, M.D. and S.A. Boyd. 1986. Complete reductive dechlorination and mineralization of pentachlorophenol by anaerobic microorganisms. *Appl. Environ. Microbiol.* 52:861.

Mikesell, M.D. and S.A. Boyd. 1988. Enhancement of pentachlorophenol degradation in soil through induced anerobiosis and bioaugmentation with anaerobic sewage sludge. *Environ Sci. Technol.* 22:1411.

Miller, R.N. 1990. A field scale investigation of enhanced petroleum hydrocarbon biodegradation in the vadose zone combining soil venting as an oxygen source with moisture and nutrient additions. Ph.D. dissertation. Utah State University. Logan, Utah.

Mohn, W.W. and J. M. Tiedje. 1991. Evidence for chemisomotic coupling of reductive dechlorination and ATP synthesis in *Desulfomonile tiedjei* strain DCB-1. In *Abst. Ann. Mtg. Am. Soc. Microbiol.* Abstract No. K4.

Mohn, W.W. and J.M. Tiedje. 1992. Microbial reductive dehalogenation. *Microbiol. Rev.* 56:482.

Montgomery, J.H. and L.M. Wilkins 1990. *Groundwater chemical desk reference.* Volume 1. New York: Lewis Publishers.

Moore, J.W. and S. Ramamoorthy. 1984. *Heavy metals in natural waters: applied monitoring and impact assessment*, 260 pp. New York: Springer-Verlag.

Nishino, S.F., J.C. Spain, L.A. Belcher, and C.D. Litchfield. 1992. Chlorobenzene degradation by bacteria isolated from contaminated groundwater. *Appl. Environ. Microbiol.* 58:1719.

Nishino, S.F., J.C. Spain and C.A. Pettigrew. 1994. Biodegradation of chlorobenzene by indigenous bacteria. *Environ. Toxicol. Chem.* 13:871-877.

Norris, R. D. and R. D. Mutch. 1991. Unpublished results.

Norris, R.D. 1993. In situ bioremediation of soils and groundwater contaminated with petroleum hydrocarbons. In *In situ bioremediation of groundwater and geological material: a review of technologies.* Office of Research and Development, R.S. Kerr Environmental Research Laboratory, U.S. Environmental Protection Agency. Ada, Okla.

Norris, R.D. and A.N. Clarke. 1991. Unpublished results.

Norris, R.D. and T.K. Subramanyam. 1992. Unpublished results.

Nyer, E.K. 1992. *Groundwater treatment technology.* 2d ed. New York: Van Nostrand Reinhold.

Nyer, E.K. and D. Ziegler. 1983. Hazardous waste destruction by submerged fixed-film biological treatment. Paper presented at the *15th Mid-Atlantic Industrial Waste Conf.*

Oldenhuis, R., R.L.J.M. Vink, D.B. Jansen, and B. Witholt. 1989. Degradation of chlorinated aliphatic hydrocarbons by *Methylosinus trichosporium* OB3b expressing soluble methane monooxygenase. *Appl. Environ. Microbiol.* 55:281.

Ostendorf, D.S. and D.H. Kampbell. 1990. Bioremediated soil venting of light hydrocarbons. *Haz. Waste Haz. Mat.* 7:219.

Overcash, M.R. and D. Pal. 1979. *Design of land treatment systems for industrial wastes-theory and practice.* Ann Arbor, Mich.: Ann Arbor Science Publications.

Phelps, T.J., D. Ringelberg, D. Hedrick, J. Davis, C.B. Fliermans, and D.C. White. 1988. Microbial biomass and activities associated with subsurface environments contaminated with chlorinated hydrocarbons. *Geomicrobiol. J.* 6:157.

Piotrowski, M.R., J.R. Doyle, D. Cosgriff, and M.C. Parsons. 1994. Bioremedial progress at the Libby, Montana, Superfund site, 240-55. In.*Appl. Biotechnol. for Site Remediation.* Ed. R.E. Hinchee, D.B. Anderson, F.B. Metting, Jr., G.D. Sayles. Boca Raton, Fla.

Piotrowski, M.R., K.E. Berryhill, and J.L. Vernor. 1993. Closed-loop in situ bioremediation of a leaking underground storage tank site. *Proceedings of Applied Bioremediation '93.* Portland Maine: Intertech.

Piotrowski, M.R. 1992. In situ designs for aquifer treatment. *Environ. Protec.* 3:36.

Piotrowski, M.R. 1991. U.S. EPA-Approved, Full-Scale Biological Treatment for Remediation of a Superfund Site in Montana. In *Hydrocarbon Contaminated Soils, Volume 1*, 433. Eds. E. Calabrese and P. Kostecki. Chelsea, Mich.: Lewis Publishers.

Piotrowski, M.R. 1989. In situ biogeochemical reduction of hydrocarbon contamination of groundwater by injecting hydrogen peroxide: a case study in a Montana aquifer contaminated by wood preservatives, 218 pp. Ph.D. Dissertation. UMI Order Number 8913768. Boston University.

Piotrowski, M.R. 1989. Improving feasibility studies. *Hazmat World* 2:42

Prokop, W.H. and H.L. Bohn. 1985 Soil bed system for control of rendering plant odors. *JAPCA* 35:1332.

Quensen, F., J.M. Tiedje and S.A. Boyd. 1988. Reductive dechlorination of polychlorinated biphenyls by anaerobic microorganisms from sediments. *Science* 242:752.

Racke, D.K. and J.R. Coats. 1990. *Enhanced biodegradation of pesticides in the environment.* Washington, D.C.: American Chemical Society.

Raymond, R.L. 1974. *Reclamation of hydrocarbon contaminated groundwaters.* U.S. Patent 3,846,290.

Raymond, R.L., R.A. Brown, R.D. Norris, and E.T. O'Neill. 1986. *Stimulation of biooxidation processes in subterranean formations.* U. S. Patent 4,588,506.

Raymond, R.L., V.W. Jamison, and J.O. Hudson. 1975. Beneficial stimulation of bacterial activity in groundwaters containing petroleum products. API Publication No. 4427. Washington, D.C.: American Petroleum Institute.

Raymond, R.L., V.W. Jamison, J.O. Hudson, and R.E. Mitchell. 1978. Field application of subsurface biodegradation of gasoline in a sand formation. API Publication No. 4430. Washington, D.C.: American Petroleum Institute.

Reed and Associates, Inc. 1988. Client Report.

Reineke, W. and H.J. Knackmuss. 1984. Microbial metabolism of haloaromatics: isolation and properties of a chlorobenzene-degrading bacterium. *Appl. Environ. Microbiol.* 47:395.

Reineke, W. and H.J. Knackmuss. 1988. Microbial degradation of haloaromatics. *Ann Rev Microbiol.* 42:263.

Rifai, H. S., P. B. Bedient, J. T. Wilson, K. M. Miller, and J. M. Armstrong. 1988. Biodegradation at an aviation fuel spill site. *ASCE J. Environ. Engr.*

Roberts, D.J., S.B. Funk and R.A. Korus. 1992. Intermediary metabolism during anaerobic degradation of TNT from munitions-contaminated soil. In *Abst. Ann. Mtg. Amer. Soc. Microbiol.* Abstract No. Q138.

Salanitro, J.P., L.A. Diaz, M.P. Williams and H.L. Wisniewski. 1994. Isolation of a bacterial culture degrading methyl t-butyl ether. *Appl. Environ. Microbiol.* 60 (in press).

Sander, P., R.M. Wittich, P. Fortnagel, H. Wilkes and F. Wittko. 1991. Degradation of 1,2,4-trichloro-and 1,2,4,5-tetrachlorobenzene by *Pseudomonas* strains. *Appl. Environ. Microbiol.* 57:1430.

Schlesinger, W.H. 1991. *Biogeochemistry: an analysis of global change*. San Diego, Calif: Academic Press, Inc.

Schraa, G., M.L. Boone, M.S.M. Jetten, A.R.W. Van Neerven, P.J.Colberg and A.J.Zehnder. 1986. Degradation of 1,4-dichlorobenzene by *Alcaligenes* sp. strain A175. *Appl. Environ. Microbiol.* 53:1374.

Schroeder, J.D., N.D. Rosenberg, E.P. Barnes-Smith, and S.R. Booth. 1992. *In situ air stripping: cost effectiveness of a remediation technology field tested at the Savannah River integrated demonstration site*, 52 pp. Department of Energy Document LA-UR-92-1927.

Schuring, J.R., Jr. 1993. Personal communication with John Cunningham at New Jersey Institute of Technology, Department of Civil and Environmental Engineering, Newark, NJ.

Semmens, M.J., T. Ahmed, and M.A. Voss. 1991. Field tests on a bubbleless membrane aerator. In *Air-Water Mass Transfer: Selected Papers from the Second International Symposium on Gas Transfer at Water Surfaces*, 694. New York, NY: American Society of Civil Engineers.

Semprini, L., P.V. Roberts, G.D. Hopkins, and P.L. McCarty. 1990. A field evaluation of in situ biodegradation of chlorinated ethenes: part 2, results of biostimulation and biotransformation experiments. *Ground Water* 28:715.

Senior, E. and M.T.M. Balba. 1984. The use of single stage and multi-stage fermenters to study the metabolism of xenobiotic and naturally occurring molecules by interacting microbial associations. In *Microbiological methods for environmental biotechnology*, 275. Ed. J.M. Grainger and J.M. Lynch. Society for Applied Bacteriology. Orlando, Fla.: Academic Press, Inc.

Sheehan, P.J., R.W. Schneiter, T.K.G. Mohr, and R.M. Gersberg. 1988. Bioreclamation of gasoline contaminated groundwater without oxygen addition. In *Proc. 2nd Natl. Outdoor Action Conf. on Aquifer Restoration, Ground Water Monitoring and Geophysical Methods,* 193. Dublin, Ohio: National Water Well Association.

Shelton, D.R., and C.J. Somich. 1988. Isolation and characterization of coumaphos-metabolizing bacteria from cattle dip. *Appl. Environ. Microbiol.* 54:2566.

Shields M.S. 1991. Construction of a *Pseudomonas cepacia* strain constitutive for the degradation of trichloroethylene and its evaluation for field and bioreactor conditions. In *Abstr. Ann. Mtg. Amer. Soc. Microbiol.* Abstract No. K8.

Shields, M.S., S.O. Montgomery, P.J. Chapman, and S.M. Cuskey, P.H. Pritchard 1989. Novel pathway of toluene catabolism in the trichloroethylene degrading bacterium G4. *Appl. Environ. Microbiol.* 55:1624.

Sims, J.L., R.C. Sims and J.E. Matthews. 1989. *Bioremediation of contaminated surface soils*. EPA/600/9-89/073. Robert S. Kerr Environmental Research Laboratory, U.S. Environmental Protection Agency. Ada, Okla.

Sinclair, J.L. and W.C. Ghiorse. 1989. Distribution of aerobic bacteria, protozoa, algae, and fungi in deep subsurface sediments. *Geomicrobiol. J.* 7:15.

Smolenski, W.J. and J.M. Suflita. 1987. Biodegradation of cresol isomers in anoxic environments. *Appl. Environ. Microbiol.* 53:710.

Spain, J.C. 1990. Metabolic pathways for biodegradation of chlorobenzenes. In *Pseudomonas: biotransformations, pathogenesis, and evolving biotechnology,* 197. Ed. S. Silver, A.M. Chakrabarty, B. Iglewski and S. Kaplan. Washington, D.C.: American Society for Microbiology.

Spain, J.C., and D.T. Gibson. 1991. Pathway for biodegradation of *p*-nitrophenol in a *Moraxella* sp. *Appl. Environ. Microbiol.* 57:812.

Spain, J.C., and S.F. Nishino. 1987. Degradation 1,4-dichlorobenzene by a *Pseudomonas* sp. *Appl. Environ. Microbiol.* 53:1010.

Spain, J.C., J.D. Milligan, D.C. Downey, and J.K. Slaughter. 1989. Excessive baterial decomposition of H_2O_2 during enchanced biodegradation. *Ground Water* 27:163-7.

Spain, J.C., O. Wyss, and D.T. Gibson. 1979. Enzymatic oxidation of *p*-nitrophenol. *Biochem. Biophys. Res. Commun.* 88:634.

Spanggord, R.J., J.C. Spain, S.F. Nishino, and K.E. Mortelmans. 1991. Biodegradation of 2,4-dinitrotoluene by a *Pseudomonas* sp. *Appl. Environ. Microbiol.* 57:3200.

Standard Methods for the Examination of Water and Wastewater. 1985. 16th ed. American Public Health Association, Washington, D.C.

Steiert, J.G., and R.L. Crawford. 1986. Catabolism of pentachlorophenol by a *Flavobacterium. Biochem. Biophys. Res. Commun.* 141:825.

Stinson, M.K., H.S. Skovronek, and T.J. Chresand. 1991. EPA SITE Demonstration of BioTrol aqueous treatment system. *J. Air Waste Manage.* 41:228-33.

Stotzky, G. 1986. Influences of soil mineral colloids on metabolic processes; growth, adhesion and ecology of microbes and viruses. In *Interaction of soil minerals with natural organics and microbes,* 305. Ed. P.M. Haung and M. Schnitzer. Madison, Wisc.: Soil Science Society of America.

Stroo, H.F. 1992. Biotechnology and hazardous waste treatment. *J. Environ. Quality* 21:167-75.

Stroo, H.F. 1991. Biological treatment of petroleum sludges in liquid/solids contact reactors. *Environ. Waste Manage. World* 3(9): 10.

Stroo, H.F. 1989. Biological treatment of petroleum sludges in liquid/solids contact reactors. *Environ. Waste Management World* 3:10.

Strubel, V., K.H. Engesser, P. Fischer, and H.J. Knackmuss. 1991. 3-(2-Hydroxyphenyl) catechol as substrate for proximal meta ring cleavage in dibenzofuran degradation by *Brevibacterium* sp. strain DPO 1361. *J. Bacteriol.* 173:1932.

Struijs, J. and J.E. Rogers. 1989. Reductive dehalogenation of dicholoranilines by anaerobic microorganisms in fresh and dichlorophenol-acclimated pond sed. *Appl. Environ. Microbiol.* 55:2527.

Struttman, T. and R. Holderman. 1992. Streamlined soil vapor extraction: remediate rather that investigate. In *Proc. 6th Natl Outdoor Action Conf,* 573. Las Vegas. May 11-13. Dublin, Ohio: National Ground Water Association.

Stucki, G. and M. Alexander. 1987. Role of dissolution rate and solubility in biodegradation of aromatic compounds. *Appl. Environ. Microbiol.* 53:292-297.

Suflita, J.M. 1989. Microbial ecology and pollutant biodegradation in subsurface ecosystems. In *Transport and fate of contaminants in the subsurface,* 67. EPA/625/4-89/019. U.S. Environmental Protection Agency, Center for Environmental Research, Cincinnati, Ohio, and R.S. Kerr Environmental Research Laboratory, Ada, Okla.

Texas Research Institute. 1982. *Enhancing the microbial degradation of underground gasoline by increasing available oxygen.* API Publication No. 4428. Washington, D.C.: American Petroleum Institute.

Thierrin, J., G.B. Davis, C. Barber, and T.R. Power. 1992b. Use of deuterated organic compounds as groundwater tracers for determination of natural degradation rates within a contaminated zone. In *6th Intl. Symp. on Water Tracing,* Karlsruhe Germany, Sept. 21-26. (in press).

Thierrin, J., G.B. Davis, C. Barber, B.M. Patterson, F. Pribac, T.R. Power, and M. Lambert. 1992a. Natural degradation rates of BTEX compounds and naphthalene in a sulfate reducing groundwater environment. In *In-Situ Bioremediation Symp. '92.* Niagara-On-The-Lake, Ontario, Canada. September 20-24. (in press).

Thomas, J. M., H. J. Marlow, C. H. Ward and R. L. Raymond. 1992. Hydrogeologic considerations for in situ biorestoration. In *Fate of pesticides in the environment,* 211. Ed. J.L. Schnoor. New York: John Wiley & Sons, Inc..

Thomas, J.M. and C.H. Ward. 1989. In situ biorestoration of contaminants in the subsurface. *Environ. Sci. Technol.* 23:760.

Thomas, J.M. and C.H. Ward. 1992. Subsurface microbial ecology and bioremediation. *J. Haz. Mat.* 32:179.

Thomas, J.M., C.H. Ward, R.L. Raymond, J.T. Wilson, and R.C. Loehr. 1992. Bioremediation. In *Encyclopedia of microbiology,* 369. Volume 1. Ed. J. Lederberg. San Diego, Calif.: Academic Press, Inc.

Thomas, J.M., M.D. Lee, and C.H. Ward. 1987. Use of groundwater in assessment of biodegradation potential in the subsurface. *Environ. Toxicol. Chem.* 6:607.

Thompson-Eagle, E.T. and W.T. Frankenberger, Jr. 1990. Volatilization of selenium from agricultural evaporation water. *J. Environ. Qual.* 19:125.

Thornton, J.S. and W.L. Wootan. 1982. Venting for the removal of hydrocarbon vapors from gasoline contaminated soil. *J. Environ. Sci. Health* A17:31.

Topp, E. and M.H. Akhtar. 1991. Identification and characterization of a *Pseudomonas* strain capable of metabolizing phenoxybenzoates. *Appl. Environ. Microbiol.* 57:1294.

Tsien, H.C., G.A. Brusseau, R.S. Hanson, and L.P. Wackett. 1989. Biodegradation of trichloroethylene by *Methylosinus trichosporium* OB3b. *Appl. Environ. Microbiol.* 55:3155.

Twenter, F. R., T. R. Cummings, and N. G. Grannemann. 1985. Ground Water Contamination in East Bay Township, Michigan. Report 85-4064 U.S. Water Resources Investigations. Geological Survey.

Urlings, L.G.C.M., H.B.R.J. van Vree, and W. van der Galien. 1990. Application of biotechnology in soil remediation. *Envirotech Vienna* 238.

US EPA. 1986. *Permit guidance manual on hazardous waste land treatment demonstrations.* Office of Solid Waste. Washington, D.C.

US EPA. 1987. *Test methods for evaluating solid waste, physical/chemcial metnods.* 3rd ed. Office of Solid Waste and Emergency Response. US EPA, SW-846. Washington, D.C.

US EPA. 1988. Appendix C: Biological treatment technologies, 103. *Technology Screening Guide for Treatment of CERCLA Soils and Sludges.* EPA/540/2-88/004. Sept.

US EPA. 1990. *Summary of treatment technology effectiveness for contaminated soil.* Office of Emergency and Remedial Response. Washington, D.C.

Valo, R. and M. Salkinoja-Salonen. 1986. Bioreclamation of chlorophenol-contaminated soil by composting. *Appl. Microbiol. Biotechnol.* 25:68.

Van der Meer, J.R., W. Roelofsen, G. Schraa, and A.J.B. Zehnder. 1987. Degradation of low concentrations of diclorobenzenes and 1,2,4-trichlorobenzene by *Pseudomonas* sp. strain P51 in nonsterile soil columns. *FEMS Microbiol. Ecol.* 45:333.

van den Wijngaard, A.J., K.W.H.J. van der Kamp, J. van der Ploeg, F. Pries, B. Kazemier, and D.B. Janssen. 1992. Degradation of 1,2-dichloroethane by *Ancyclobacter aquaticus* and other facultative methylotrophs. *Appl. Environ. Microbiol.* 58:976.

van Eyk, J. and C. Vreeken. 1989a. Model of petroleum mineralization response to soil aeration to aid in site-specific, in situ biological remediation. In *Groundwater contamination: use of models in decision-making. Proc. Intl Conf. on Groundwater Contamination,* 365. Ed. Jousma et al. Kluwer Bost/London.

van Eyk, J. and C. Vreeken. 1989b. Venting-mediated removal of diesel oil from subsurface soil strata as a result of stimulated evaporation and enhanced biodegradation. In *Hazardous waste and contaminated sites,* 475. Envirotech Vienna, Volume 2, Session 3, ISBN 389 432-00905. Westarp Wiss., Essen.

Vandegrift, S. A. and D. H. Kampbell. 1988. Gas chromatic determination of aviation gas and JP-4 jet fuel in subsurface core samples. *J. Chrom. Sci.*

Verschueren, K. 1983. *Handbook of environmental data on organic chemicals.* 2d ed. New York: VanNostrand Reinhold.

Vesper, S.J., M. Narayanaswamy, L.C. Murdoch, and W.J. Davis-Hoover. 1994. Hydraulic fracturing to enhance in situ bioreclamation of subsurface soils. In *Applied Biotechnology for Site Remediation,* 36. Eds. R.E. Hinchee, D.B. Anderson, F.B. Metting, Jr., and G.D. Sayles. Boca Raton, Fla.: Lewis Publishers.

Vogel, T.M. and P.L McCarty. 1985. Biotransformation of tetrachloroethylene to trichlorethylene, vinyl chloride, and carbon dioxide under methanogenic conditions. *Appl. Environ. Microbiol.* 49:1,080.

Vogel, T.M., C.S. Criddle, and P.L. McCarty. 1987. Transformations of halogenated aliphatic compounds. *Environ. Sci. Technol.* 21:722.

Voronin, L.M. 1991. Simulation of ground-water flow at Picatinny Arsenal, New Jersey. In *Proc. Technical Meeting: U.S. Geological Survey Toxic Substances Hydrology Program,* 713. U.S. Geological Survey Water Resources Investigations Report 91-4034. Monterey, Calif.

Wackett, L.P., G.A. Brusseau, S.R. Householder, and R.S. Hanson. 1989. Survey of microbial oxygenases: trichloroethylene degradation by propane oxidizing bacteria. *Appl. Environ. Microbiol.* 55:2960.

Ward, C. H., J. M. Thomas, S. Fiorenza, H. S. Rifai, P. B. Bedient, J. T. Wilson, and R. L. Raymond. 1989. In situ bioremediation of subsurface material and groundwater contamnated with avaitin fuel: Traverse City, Michigan. In *Proc. 1989 Air and Waste Management/Environmnetal Protection Agency International Symposium,* 83-96. Cincinnati. February 1989.

Williams, R.T. and R.S. Ziegenfuss. 1989. Composting of explosives and propellant contaminated sediments. In *Proc. 3rd Intl. Conf. on New Frontiers for Hazardous Waste Management,* 204. Pittsburgh.

Williams, R.T., P.S. Ziegenfuss, and W.E. Sisk. 1992. Composting of explosives and propellant contaminated soils under thermophilic and mesophilic conditions. *J. Indust. Microbiol.* 9:137.

Wilson, B.H. 1988. Biotransformation of chlorinated hydrocarbons and alkylbenzenes in aquifer material from the Picatinny Arsenal, New Jersey. In *Proc. Technical Meeting: U.S. Geological Survey Toxic Substances Hydrology Program.* Report 88-4220. U.S. Geological Survey Water Resources Investigations. Phoenix.

Wilson, B.H., J.T. Wilson, D.H. Kampbell, B.E. Bledsoe, and J.M. Armstrong. 1990. Biotransformation of monoaromatic and chlorinated hydrocarbons at an aviation gasoline spill site. *Geomicrobiol. J.* 8:225.

Wilson, B.H., T.A. Ehlke, T.E. Imbrigiotta, and J.T. Wilson. 1991. Reductive dechlorination of trichloroethylene in anoxic aquifer material from Picatinny Arsenal, New Jersey. In *Proc. Technical Meeting: U.S. Geological Survey Toxic Substances Hydrology Program,* 704. Report 91-4034. U.S. Geological Survey Water Resources Investigations. Monterey, Calif.

Wilson, D.J. 1992. Private Communication.

Wilson, D.J., A.N. Clarke, and J.H. Clarke. 1988. Soil clean up by in situ aeration I, mathematical modeling. *Sep. Sci. Technol.* 23:991.

Wilson, D.J., S. Kayano, R.D.Mutch, Jr., and A.N. Clarke. 1992. Groundwater clean-up by in-situ sparging: mathematical modeling. *Sep. Sci. Technol.* 27:1023.

Wilson, J. T., J. M. Armstrong, and H. Rifai. 1993. A full-scale field demonstration on the use of hydrogen peroxide for in situ bioremediation of an aviation gasoline-contaminated aquifer, 333-58. In *Bioremediation: Field Experiences*. Ed. P. E. Flathman, D. E. Jerger, and J. H. Exner. Chelsa, Mich.: Lewis Publishers.

Wilson, J.T. and B.H. Wilson. 1985. Biotransformation of trichloroethylene in soil. *Appl. Environ. Microbiol.* 49:242.

Wilson, J.T. and C.H. Ward. 1986. Opportunities for bioremediation of aquifers contaminated with petroleum hydrocarbons. *J. Ind. Microbiol.* 27:109.

Wilson, J.T., D.H. Kampbell, and J. Armstrong. 1993. Natural bioreclamation of alkylbenzenes (BTEX) from a gasoline spill in methanogenic groundwater. Paper presented at *2nd Intl. Symp.: In Situ and On-site Bioreclamation*. San Diego, Calif.. April 5-8.

Wilson, J.T., J.F. McNabb, J. Cochran, T. Wang, M.B. Tomson, and P.B. Bedient. 1985. Influence of microbial adaptation on the fate of organic pollutants in groundwater. *Environ. Toxicol. Chem.* 4:721.

Winter, R.B., K.M. Yen, and B.D. Ensley. 1989. Efficient degradation of trichloroethylene by a recombinant *Escherichia coli. Bio/Tech.* 7:282.

Zeyer, J. and P.C. Kearney. 1984. Degradation of *o*-nitrophenol and *m*-nitrophenol by a *Pseudomonas putida. J. Agric.Food Chem.* 32:238.

Zeyer, J., A. Wasserfallen, and K.N. Timmis. 1985. Microbial mineralization of ring-substituted anilines through an ortho-cleavage pathway. *Appl. Environ. Microbiol.* 50:447.

Zylstra, G.J., L.P. Wackett, and D.T. Gibson. 1989. Trichloroethylene degradation by *Escherichia coli* containing the cloned *Pseudomonas putida* F1 toluene dioxygenase genes. *Appl. Environ. Microbiol.* 55:3162.